[英国] 詹姆斯·宾尼 著 杨晨 译

牛津通识读本·天体物理学
Astrophysics
A Very Short Introduction

译林出版社

图书在版编目（CIP）数据

天体物理学 /（英）詹姆斯·宾尼（James Binney）著；杨晨译. -- 南京：译林出版社，2025.1.
(牛津通识读本). -- ISBN 978-7-5753-0449-8

Ⅰ. P14

中国国家版本馆CIP数据核字第2024P9S888号

Astrophysics: A Very Short Introduction, First Edition by James Binney
Copyright © James Binney 2016
Astrophysics: A Very Short Introduction, First Edition was originally published in English in 2016.
This licensed edition is published by arrangement with Oxford University Press.
Yilin Press, Ltd is solely responsible for this bilingual edition from the original work and Oxford University Press shall have no liability for any errors, omissions or inaccuracies or ambiguities in such bilingual edition or for any losses caused by reliance thereon.
Chinese and English edition copyright © 2025 by Yilin Press, Ltd
All rights reserved.

著作权合同登记号　图字：10-2018-429号

天体物理学　　［英］詹姆斯·宾尼　/　著　杨　晨　/　译

责任编辑　许　丹
装帧设计　景秋萍
校　　对　施雨嘉
责任印制　董　虎

原文出版　Oxford University Press, 2016
出版发行　译林出版社
地　　址　南京市湖南路1号A楼
邮　　箱　yilin@yilin.com
网　　址　www.yilin.com
市场热线　025-86633278
排　　版　南京展望文化发展有限公司
印　　刷　江苏扬中印刷有限公司
开　　本　890毫米×1260毫米　1/32
印　　张　9.875
插　　页　4
版　　次　2025年1月第1版
印　　次　2025年1月第1次印刷
书　　号　ISBN 978-7-5753-0449-8
定　　价　39.00元

版权所有·侵权必究

译林版图书若有印装错误可向出版社调换。质量热线：025-83658316

序 言

赵 刚

16世纪中叶，哥白尼创立的日心说标志着近代天文学的肇始。17世纪，望远镜的发明为天文学带来了革命性变化，牛顿提出的万有引力定律则奠定了天体力学基础。19世纪，光谱学被应用于天文学，通过分析天体光谱以研究其物理和化学性质，这成为天体物理学的开端。20世纪，相对论和量子力学创立，为天体物理学带来新的发展方向。恒星演化理论的建立、射电天文学的兴起和空间天文学的发展为人类提供了前所未有的全波段宇宙图景。21世纪，引力波的直接探测则开启了多信使天文学的新纪元。

对于想要了解天体物理学的人来说，这本《天体物理学》无疑是一本很好的入门书。作者詹姆斯·宾尼（James Binney）是一位杰出的理论天体物理学家，同时他还是牛津大学荣休教授、英国皇家学会院士、美国科学院外籍院士，以银河系和河外星系理论研究而闻名遐迩。宾尼教授的学术成就广受认可，特别是他撰写的两本星系方向研究专著《星系天文学》（2004）和

《星系动力学》(2005),在天文界颇为流行,其中前者由我和我的学生联合翻译出版,后者由上海天文台宋国玄先生翻译出版。这两本书长期以来一直被用作天体物理领域的教科书或参考书,而这本《天体物理学》是牛津大学 Very Short Introductions 系列丛书的经典之作,为广大读者开启了一扇通往天体物理学奇妙世界的大门。

詹姆斯·宾尼教授在这本书中展示了天体物理学领域在过去一个世纪里是如何迅速发展的,天体物理学如何利用地球上得出的物理定律去理解宇宙中那些遥不可及的天体和天体现象。作为一门综合性极强的学科,天体物理学融合了物理学、天文学、数学、化学等多学科的知识与方法。宾尼教授避开了繁杂的数学推导与专业术语的堆砌,以一种通俗易懂、生动形象的方式,向读者介绍了天体物理学的核心概念、关键理论以及重大发现。这本书从有关天文的一些重大思想讲起,过渡到星际气体、恒星、吸积、行星、星系等主题,从中描绘出宇宙的起源与命运的宏大场景。宾尼教授以清晰的逻辑、鲜活的实例、精练的篇幅和深入浅出的笔触,让即使没有深厚物理学背景的人也能够轻松理解,让读者仿佛置身于广袤无垠的宇宙之中,目睹着不可思议的天文奇观。

作为宾尼教授的同行和老朋友,我曾有幸收到他亲笔签名的此书的英文原版。如今,欣闻这部著作的双语版《天体物理学》即将问世,我感到格外高兴,它将让更多中文世界的读者有机会领略天体物理学的丰富内涵与持久魅力。

《天体物理学》双语版的出版,为中文读者提供了一本高质量的天体物理学科普读物。本书的译者杨晨是北京理工大学硕士，热爱翻译。我也阅读了"牛津通识读本"系列中他已经出版的另外两本译著《物理学》和《元素周期表》，行文流畅，皆为佳作，可以看出其具有很扎实的理科背景和深厚的翻译功力。由此，各位读者可以对这本《天体物理学》译作充满期待。

  本书既可以作为普通读者了解天体物理学基础知识的入门书籍，也可以作为相关专业学生和科研人员拓展视野、启发思维的有益参考。希望广大读者跟随宾尼教授的视角，去感受天体物理学的无穷魅力，并燃起对宇宙探索的热情。通过阅读这本书，相信你一定可以体会到"开卷有益，掩卷有味"的读书愉悦。

2024年11月
于中国科学院国家天文台

目 录

第一章　重大的思想　1

第二章　恒星间的气体　10

第三章　恒　星　20

第四章　吸　积　45

第五章　行星系　64

第六章　相对论天体物理学　78

第七章　星　系　104

第八章　总览全局　131

　　　　索　引　135

　　　　英文原义　147

第一章

重大的思想

天地同规律

牛顿之前只有天文学,之后才有了天体物理学。传说他在自己的伍尔索普果园里看到苹果落下,顿时灵光一闪,想到月球也应该和苹果一样落下。如果这是真的,那么天体物理学便在此刻诞生了。也就是说,月球等天体并不像牛顿的前辈设想的那样,按照神明定下的轨迹在天空滑行,它们与一个明天就可能被鸟兽啃去一半、让人不屑一顾的苹果一样,都遵循着相同的物理定律。

这个思想的力量,在于我们可以用实验室中提炼的物理定律解释远在宇宙深处的天体。于是,牛顿的思想使我们在思维中旅行,穿过宇宙那超乎想象的广袤空间,观察遥远星系中心的巨大黑洞——而如今即使用上射电望远镜,也只能收到微弱的信号。

牛顿还从另一个重要方面奠定了天体物理学的基础:我们可

以通过具有恰当定义的物理定律,得到精确的定量预测。因此,他不仅能对已有的观测做出一致的物理解释,还能**预测**未来可能观测到的结果。为此,他必须发明一种新的数学——微积分,用它的语言概括物理定律。牛顿时代之后,大部分物理定律便采用了微分方程的形式,它通过描述函数的变化率来确定函数。微分方程包含了给定物理情境的一切情况,我们需要根据初始条件找到表示特定情形的函数。例如,手枪打出的子弹的轨迹是牛顿方程 $m\,\mathrm{d}v/\mathrm{d}t = \mathbf{F}$ 的解。这个方程通常简化为 $f = ma$,将速度($\mathbf{v}$)的变化率(即加速度)与力($\mathbf{F}$)联系了起来。牛顿方程适用于一切子弹和下落的苹果,也适用于月球,是个普适方程。月球、子弹、苹果之所以有不同的轨迹,是因为它们的初始条件不同:月球远离地心,运动速度非常快;子弹在地表附近射出,速度慢得多;苹果也在地表附近运动,但它一开始是静止的。不同的初始条件应用到相同的普适方程中,就得到了三条完全不同的轨迹。于是,牛顿发明的数学变身为有力的工具,我们由此确定是什么让不同的现象具有共性,又是什么让它们有所区别。

其中必有关联

詹姆斯·克拉克·麦克斯韦是爱丁堡一个富裕律师的独子。他很早就显露出数学和物理的高超天赋,在气体和热理论,以及土星环的动力学分析等领域做出了巨大的贡献。但他最大的成就,是通过纯思维的方式拓展了电磁学的定律。他设想了一个特别的实验装置,其中有交流电在电路中来回流动。电路里有一个电容,这是一种由两块金属板组成的元件,中间夹着一层很薄的绝缘体,原则上可以是真空。电流流到一块金属板上,使它带正电，从另一块金属板流出，使它带负电。麦克斯韦运用安德烈-玛丽·安培提出的法则计算这个电路产生的磁场。他在1865年证明，如果以不同的方式运用安培的法则，就会得出全然不同的结果，除非电容金属板之间的绝缘体中也有电流通过。这个结果使麦克斯韦提出假设，认为随时间变化的电场会产生"位移电流"。从数学上看，原先的微分方程表达了传统电流与它生成的磁场的关联，而这假想的电流在其中添加了额外的一项。

这个额外的项带来了惊人的影响，它使得电场和磁场能够相互维持，而不需要电荷参与其中。在此之前，电场围绕在带电物体周围，磁场则围绕在通电线圈周围。但有了这个额外项后，随时间变化的电场能够生成一个随时间变化的磁场。迈克尔·法拉第已经证明，这样的磁场可以生成随时间变化的电场。因此，在不需要任何电荷参与的情况下，这个磁场又可以重新生成原先的电场！这个惊人的结论正确吗？方程中的额外项是否是愚蠢的错误？

麦克斯韦计算了电场和磁场的耦合振荡在真空中传播的速度，结果与光速测量值在实验误差内吻合。于是麦克斯韦得出结论说，他的额外项是正确的，光是电场和磁场相互维持的振荡。因为光的波长非常短（约为0.0 005毫米），所以振荡频率一定很高。频率较低的振荡对应较长的波长。1886年，海因里希·赫兹发射并检测到了这样的"射电"波。

因此，麦克斯韦重新解释了一个古老的现象——光。他将传统的物理学定律应用于思想实验之中，然后论证这些定律需要修正，从而使整个理论前后**一致**，并由此预言了一种全新现象。这一步，石破天惊。

永永远远

我们相信物理定律永远正确：有确凿证据证明，它们自138亿年前宇宙诞生后的一分钟起便是如此。宇宙从爆炸的火球经历寒冷、黑暗的时期，创生出第一批恒星与星系（现在正由巨型望远镜研究），其间的物理定律始终是对的，而且至今依然正确。

虽然物理定律在过去的138亿年间一直保持着稳定，但宇宙却已然改头换面。这里再次体现了牛顿式物理定律与物理定律所描述的现象的区别：物理定律体现为微分方程，它在任何时刻、任何地点都是对的；物理现象则会因为方程需要的初始条件急剧变化而完全改变。

因为物理定律在宇宙中的每个部分都有效，所以我们可以在头脑中畅游遥远的星系；因为物理定律在任何时间都有效，所以我们可以在头脑中回溯最初的时刻。物理定律这种普遍、永恒不变的性质让我们在想象中成为时空的旅行者。

将物理定律应用到地球之外的一切事物，这便是天体物理学。因此，它是其他科学的晚辈，但在规模上傲视群雄。

太初有道

宇宙瞬息万变，而物理定律永恒，它们在宇宙诞生前便存在，是它们构建了宇宙。实验过程不可能每天都一样，因为现实世界中的万物都在变化。今天比昨天凉一些，多少**会**让实验过程发生变化。地球的磁场在不停地转变方向，多少会影响实验。太阳越来越老，越来越亮；月球在逐渐漂离地球——这些事实也多少会影响实验。现实世界没有固定不变的事物，但在物理学家的思维世界中，定律永远为真、永恒不变。这种永恒不是偶然也不是妄念，而是意志的行为。物理学家只有将现象回溯成永远为真的定律，才会觉得自己真正理解了现象。

如果将实验设备打包运送到另一个国家、另一个纬度，实验过程会有些不一样，因为当地的地磁场、引力场与原先不同，气温可能更高或更低，穿过实验室的宇宙射线流量也不同。但物理定律依旧完全不变。同样，认为物理定律在这里那里、在一切地点都相同，也是意志的行为：只有将实验在新老地点的差异归因为某种环境上的差异，并由此通过不可变且普适的物理定律得到这不同的结果，我们才肯罢休。

坚持用无所不在且永远正确的定律解释各种现象，不仅可以使我们穿越宇宙，在空间与时间中漫游，回到最遥远的时代，还为我们装备了三件强大的武器，随我们一同旅行。它们是能量、动量和角动量。

1915年，埃米·诺特证明了一个重要结论：如果掌控系统运动的定律在系统平移或旋转后保持不变，那么便可以根据系统平移或转动的位置和速度求出一个始终保持不变的量。我们称这样的系统拥有一个"守恒量"。由每处的定律都相同，得到的守恒量是动量；由系统不论朝向东西南北还是别的什么方向都没有变化，得到的守恒量是角动量。诺特定理有一个推论：如果动力学规律在任何时候都相同，就存在另一个守恒量——能量。因此，由物理定律的普适性和不变本质可以得到动量、角动量和能量这三大守恒量。若要理解非常遥远或非常久远的系统，这三个量的守恒性质将提供很大的帮助。

1930年，沃尔夫冈·泡利设想了一种名为**中微子**的粒子，它会在核反应中带走动量和能量。他之所以做出这个假设，是因为实验证据清晰地表明，核反应过程中的能量和动量不守恒。于是他猜测存在一种不可见的粒子，由此保证能量和动量**依然**守恒。之后大约三十年里，中微子不过是纯粹的猜测，但在1956年，它们终于被发现了。中微子难以探测，因为它们与其他物质的反应截面很小，只有大约 10^{-46} 平方米。按照经典物理，其他粒子要想与中微子发生碰撞，就必须经过中微子附近大约 $\sqrt{10^{-46}}$ 米即 10^{-23} 米以内的区域，这个距离是质子直径的一亿分之一。其实根据量子力学，这么精确的定位没有任何意义，所以中微子反应截面极小的真正意义是，与它们发生相互作用的可能性极低。尽管如此，中微子在构建宇宙的过程中依然发挥了重要作用。

天空之上新事多

本书的开头，牛顿将月球拉入凡尘，使它遵从一般的运动定律。但在1930年代，瑞士怪才天文学家弗里茨·兹威基又为天空恢复了些许地位。他说："只要能发生的，便一定会发生。"这句话的意思是，**任何**遵从物理学定律的事情，都一定会在宇宙某处发生，而且只要有对应的仪器和一点点运气，我们就能**看到**它们。兹威基原理表明，努力思考原则上可能出现的奇怪对象和怪异事件，这并非徒劳之举。如果你的物理学得不错，就可以算出这些对象和事件中有哪些可以观测，甚至还能估算它们发生的频率。然后，你就可以催促观测人员去寻找它们了。

这个思路的经典案例是白矮星的发现。1930年，苏布拉马尼扬·钱德拉塞卡登上孟买开往南安普顿的远洋渡轮，准备去剑桥大学学习。途中，他想知道当时新兴但充满争议的量子力学对恒星有什么影响,于是证明了太阳等恒星在能源耗尽后会缩小到很小的体积(太阳会收缩成地球这么大)。届时,维持这个密度极大的星体的形态,使它不会在自己强大的引力下继续塌缩的压力,就是纯粹的量子力学产物:尽管星体已经冷却,但它的电子却在以接近光速的速度飞驰,因为如果不飞这么快,其中能量最低的电子就会违背海森堡的不确定性原理。这个原理要求电子在位置比较确定的情况下,具有非常不确定的动量。再者,泡利不相容原理禁止两个电子占据相同的量子态,所以大部分电子必须处在相当高的能态,因为刚好不违背海森堡原理的能态已经被占满了。

这个精彩的理论让钱德拉塞卡满心欢喜。他兴冲冲来到剑桥,却被英国天体物理学巨擘亚瑟·爱丁顿爵士浇了一盆冷水:爱丁顿觉得他在胡说八道,不予理会。爱丁顿既不接受兹威基原理,也不认同将量子力学这个(勉强凑合着)解释原子行为的古怪理论用在恒星上。但钱德拉塞卡是对的,太阳附近就有大量这类冷却、致密的星体,靠纯粹的量子力学效应维持着形态。

在第三章和第八章,我们还会遇到其他以独特的方式运用物理学,由此得到离奇的结果并做出成功预言的案例。兹威基原理之所以有效,是因为宇宙大而多变,其中自然进行着不计其数的实验。我们的地球固然十分有趣,但也十分有限。如果想要了解物质世界的广大,还是偶尔抬起头,看看天外吧。

谈谈单位

标准的科学单位千克、米、秒等适用于日常生活,若用于天体物理学,就得带上非常大的数。对我们而言,更方便的质量单位是太阳质量：$M_\odot = 2.00 \times 10^{30}$ kg，其中 10^{30} 是 1 后面跟 30 个 0 的简写。

考察行星系时，**天文单位**（AU）是便于使用的长度单位，它表示地球到太阳的平均距离：$1\ \text{AU} = 1.50 \times 10^{11}$ m，即 1.5 亿千米。在星系和宇宙尺度下，即使天文单位也不够大，用起来不方便。这时使用的距离单位是**秒差距**（pc），在地球上观测与太阳相对静止、距离 1 秒差距的恒星，会看到它每季度在天空移动 1 角秒（图 1）。由三角学可得 $1\ \text{pc} = 2.06 \times 10^5\ \text{AU} = 3.09 \times 10^{16}$ m。最近的恒星差不多就在 1 秒差距之外，而银河系中心距离我们 8.3×10^3 pc = 8.3 kpc。平均而言，与银河系一样亮的星系，体积大约是 10 兆立方秒差距。

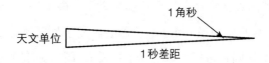

图 1　1 秒差距是使太阳与地球的距离（1 天文单位）对应 1 角秒（1/3600°）的长度

我们常以年（$1\ \text{yr} = 3.16 \times 10^7$ s）作为时间单位，尽管我们经常遇到更长的时标：恒星演化要经历几百万甚至几十亿年。因此，我们常用兆年（Myr）表示 100 万年，用吉年（Gyr）表示 10 亿年，它们的关系是 $1\ \text{Gyr} = 1000\ \text{Myr} = 10^9$ yr。

经验表明，千米每秒（km/s）是比较方便的速度单位：地球绕太阳公转的速度约为 30 千米每秒，太阳绕银河系旋转的速度约为 240 千米每秒。以 1 千米每秒的速度运动，分别要用 100 万年和 10 亿年才能走完 1 秒差距和 1 千秒差距。例如，在 10 亿年里，太阳会在围绕银河系的轨道上移动大约 240 千秒差距，由于轨道周长为$2\pi \times 8.3 \text{ kpc} = 52 \text{ kpc}$，所以太阳在10亿年里差不多能绕银河系五圈。

功率的标准单位是瓦特（W，1瓦大致等于每秒将1千克重物提升0.1米的功率）。方便天体物理学的功率是太阳的光度：$L_\odot = 3.85 \times 10^{26}$ W。公共电力一般按千瓦时计费，相当于$3.0 \times 10^{-28} L_\odot \text{yr}$。超新星爆发（见第三章的"恒星爆发"）向附近星际气体放出的能量约为$8.2 \times 10^9 L_\odot \text{yr}$。

$L_\odot \text{yr}$对天体而言是个很方便的单位，但它完全不适合原子。讨论原子和亚原子物质时，适用的能量单位是电子伏（eV）。1电子伏是将电子移动1伏电势差所需的能量，约为$10^{-53} L_\odot \text{yr}$。人眼能辨认的光子大约携带2电子伏能量，因此太阳每年会释放大约10^{53}个光子。

第二章

恒星间的气体

虽然恒星之间的空间比地球上创造的任何真空都更空,但它们也不完全是空的。太阳附近的气体平均每立方厘米含有一个原子,而每立方厘米的空气含有大约 10^{19} 个原子,因此太阳附近的区域可以称得上是 10^{-19} 巴①的超高真空。

星际吸收与红化

星际气体极为稀薄,主要由氢原子和氦原子组成,它们有多种多样的表现途径。其中最简单也最重要的是吸收恒星的光。其实,吸收光的不是气体本身,而是嵌在气体中的烟尘微粒。天文学家将它们称为**尘粒**,但其实"烟尘"(smoke)要恰当得多。我们会在第三章的"主序之后"一节看到,它们所在的气体是被某些恒星喷射出来的,就像是蜡烛燃烧时升起的白雾,或是篝火的火焰上翻滚的浓烟。

---

① 巴是压强单位,1巴 = 100千帕。——译者注(除非另有说明,本书脚注均为译者注)

自然，尘粒吸收恒星光的效率取决于尘粒的密度，因而也就取决于它们所在气体的密度。事实证明，银河系内每单位质量气体含有的尘粒质量大体上保持恒定均匀。在少数方向上，每单位面积天空中可见恒星的数量大幅下跌，因为在这些方向存在浓密的气体云，遮蔽了后面的恒星（图2）。

如果透过篝火的烟雾看太阳，太阳会显得比平时更红，因为蓝光比红光更容易被微粒吸收。所以，太阳发出的红光比蓝光更容易穿过浓烟。这种选择性吸收背后的物理原理是，天线不

图2　一处暗球状体

容易接收波长远大于自身尺寸的辐射。1960年代使用**超高频**（约为0.3千兆赫）广播信号后，电视机的天线就变得短多了。能够处理15厘米左右波长辐射的电子元件变得廉价后，移动电话也变得更小，不再带有外置的可见天线。事实表明，大多数星际尘粒的尺寸不足1微米（10^{-3}毫米），所以波长大于几微米的波不怎么会被尘粒吸收。实际上，我们可以用波长为几微米的光波直接观测浓密星际云的深处，这个波段比可见光约为0.5微米的波长大四倍左右。

因为尘粒能有效吸收蓝光和紫外线，所以透过星际云看到的恒星比云层较薄时的同类恒星更红。对比这样一对恒星的颜色，可以测定较红恒星的**红化**程度，由此确定我们与它之间的尘粒数量，进而确定气体量。星际气体最早就是用这种方法证实的。

尘粒的调节功能

尘粒在调节气体温度、密度和化学组分方面发挥着重要作用。星际空间飞驰游荡的电子和质子有时会撞入尘粒，产生冲击波使尘粒振荡，而振荡又使尘粒辐射电磁波。由此，电子和质子的一部分动能转化成了电磁波。后面会看到，即使在浓密的气体云中，它们也很容易逸散出去。于是，尘粒成了星际气体主要的冷却剂。

我们已经知道，尘粒会大量吸收恒星发出的光，尤其是蓝光和紫外线。它们自然也会像晒日光浴那样，因为吸收辐射而变热。又因为它们的质量极小，所以即使只有一个光子，也能大幅提升尘粒的温度。也就是说，一个光子就能让一颗尘粒剧烈振动。在尘粒吸收光子前因为碰撞而附着的电子或质子，会因为振动而脱离，就像是游完泳的狗将身上的水甩下来一样。如果尘粒甩出的电子和质子的速度比它们先前撞击时的速度更快，那么总体上看，尘粒会加热星际气体。因此，尘粒既可以冷却星际气体，也可以加热星际气体，这取决于射入气体的恒星光的强度。

如果恒星光比较微弱，那么尘粒在两次吸收光子之间可以积累多个质子和电子。于是，质子在尘粒表面晃动时，可能会靠得非常近，从而形成氢分子（H_2）。合成氢分子会释放能量，传递给尘粒。之后光子再加热尘粒时，氢分子有可能就此脱离。所以，尘粒提供了氢分子合成的主要途径。

尘粒还能调控其他原子的结合。星际气体包含的碳、氮、氧、硫等原子，虽然比氢原子和氦原子少得多，但丰度依然很高。如果一颗尘粒同时吸附了碳原子和氧原子，就很容易形成一氧化碳分子（CO）。如果一颗尘粒同时携带着碳原子和氮原子，就很容易形成一种毒性更大的分子——氰化氢（HCN），因为附近总会有氢原子加入。尘粒就是以这种方式调控着星际气体的化学组分。

气体的发射

关于星际气体，我们所知的大部分信息是费了很大工夫探测星际原子和分子的辐射搜集的。氢原子由一个质子和一个绕质子旋转的电子组成，它会以两种大相径庭的方式释放辐射。一种是电子自旋相对于质子自旋发生翻转：二者自旋反向的原子，能量会略高于同向的原子，所以反向原子在电子自旋翻转时会发射一个光子。这个光子的波长（21厘米）远远大于原子的大小，所以原子发射这种长波辐射的效率非常低。实际上，如果考察单个反向原子，那么它可能会保持原状上亿年，才会坍缩到能量更低的状态。所幸，星际空间的深处没有太多干扰，那里有非常多的氢原子等着完成这一跃迁。所以，如果将广播天线调到这个奇特的频率，就能接收到翻转原子的强烈信号。这种信号由弗兰克·范德赫斯特率先预言，当时他还是学生，在纳粹占领下的荷兰莱顿上学。真正探测到它，则是得益于战时的雷达研究。1951年，荷兰、澳大利亚、美国的研究团队同时宣布发现了这个信号。

探测到氢原子21厘米的辐射，使我们大幅加深了对银河系的了解，因为我们由此第一次获得了银河系旋转的清晰图像。辐射透露了发射原子的运动状态，因为对于相对原子保持静止的观察者而言，原子辐射的频率已经有了精确的测量结果。如果原子相对于观察者在运动，测到的频率也会变化：靠近观测者时频率变大，远离时变小（即多普勒效应）。

氢分子不会与能量低于紫外线的光子相互作用，而满足条件的光子很少，所以氢分子几乎不可见。这是天文学家面对的一个大问题，因为银河系的星际气体大约有一半是氢分子。而且，它们在许多方面是最重要的那一半，因为它们又冷又致密，能够转变为恒星与行星。幸好一氧化碳能很好地标示氢分子的位置。一氧化碳分子是"电偶极子"，因为电子分配不均匀，氧原子能控制更多电子，使碳原子那端带正电，氧原子那端带负电。除非气体温度极低，否则一氧化碳分子会在运动时旋转，而旋转的电偶极子会发射电磁波。这些波的频率非常精确，因为量子力学会将自转限定为一系列的离散值：分子可以无旋转

（量子数 $j = 0$），或以 1 单位（$j = 1$）、2 单位等旋转，以此类推。此外，分子每次只能将旋转状态改变 1 单位，当它从旋转状态 j 变为 $j - 1$ 时会发射一个光子，光子携带的能量正比于 j。所以，出射光子的所有频率都是由 $j = 1$ 跃迁到 $j = 0$ 的基础频率的倍数。基态光子的波长为 2.3 毫米，因为波长与频率成反比，所以由 j 跃迁到 $j - 1$ 的光子波长为 $2.3/j$ 毫米。

某分子以 j 旋转的概率取决于气体的温度。如果温度低，能量就不够多，旋转快或运动快的分子更少。而在高温时，分子的旋转和运动都会变快。因此，$j = 4$ 与 $j = 1$ 的分子数比例会随着温度上升而增大。由此，随着温度上升，与波长 2.3 毫米谱线的强度相比，波长为 2.3/4 毫米的谱线强度会变强。因此，我们可以测量多条谱线，进而确定气体的温度。

到了 1970 年代，我们终于能够探测银河系的前几条谱线，由此绘制了星际介质中更致密、温度更低区域的分布图。

我们早已绘制了邻近星系的 21 厘米和 2.3 毫米谱线图。现在，我们能够探测极远星系的一氧化碳，而且全球多国正在合作建设一座巨型射电望远镜——平方千米阵列，预计在 2020 年后绘制第一批恒星和星系形成前发射 21 厘米辐射的气体的分布图[①]。

处于基态的氢原子不会吸收可见光光子，但它会吸收能量高的紫外线光子，后者携带的能量超过了 10.2 电子伏。携带 10.2 电子伏能量的光子叫作莱曼 α 光子，它们在天文学中起着重要作用，因为它们很容易被氢原子散射——氢原子会吸收一个光子，然后再将它从另一个方向发射出去。

---

① 根据国家天文台的项目介绍和搜索到的相关信息，2023 年还没有投入使用。

能量高于13.6电子伏的光子会将氢原子的电子剥离。也就是说,它们能将氢原子电离,把它拆成一个自由电子和一个质子。随后,质子很可能会捕获路过的自由电子,同时发射一个光子。第一个出射光子大概只携带着少许能量,因为电子一开始可能只受到了轻微的束缚。但电子一旦捕获成功,就很可能会像一个在楼梯上失去了平衡的醉鬼那样,不断地"下落",在质子的电场中越陷越深,每下落一步就发射一个光子。电子落到阶梯倒数第二级时释放的光子叫作**巴耳末光子**。能量最低的巴耳末光子是Hα光子,呈现出漂亮的粉红色,在天文照片中出现在生成恒星的地方,因为这些区域的炽热恒星会将周围的气体电离,为质子、电子提供了重新结合的场所。

紫外线光子不仅对原子有巨大影响,对分子也是如此,因为它们会将分子拆解为原子,也就是会将分子**离解**。实际上,这是分子被破坏的主要方式——分子在尘粒中形成,又被紫外线光子摧毁。因此,星际气体的化学组分取决于紫外线光子破坏力与尘粒合成作用的平衡。气体的密度越高,原子与尘粒就碰撞得越频繁,合并为分子的原子占比也更大。而且,尘粒的密度越大,热恒星的紫外线光子在离解分子前就被它们吸收的比例也越大。因此,分子形式的气体比例随气体密度增大而迅速增大。

如果某个区域的气体密度变大,那里就更容易出现失控,因为密度会不断增大,几乎没有尽头。之所以会失控,是因为随着气体密度增大,附近恒星发射的紫外线光子更难不被吸收而进入气体云内部。但我们已经知道,尘粒是星际气体的主要热源,一氧化碳等分子会辐射能量。因此,紫外线光子密度下降、分子占比上升,会导致气体冷却。同密度下,更冷的气体产生的压强更低,因此气体在冷却过程中,会因为自身引力的牵引而收缩。气体密度越来越高,紫外线光子越来越少,气体越来越冷、越来越致密。这样的密度失控先是形成了图2中的暗球状体,然后便形成了恒星。

气体盘

氢原子21厘米线与一氧化碳2.3毫米线的观测结果显示,银河系中平面外围绕着一层薄薄的气体。这层气体距离银心大约4千秒差距,以接近圆周运动的速度移动。一氧化碳比氢原子更向中心集中,也更不均匀。它也更集中于中平面,大多数在中平面大约40秒差距的范围内,而氢原子则分布在大约100秒差距内。

有时,星际物质会被巨大的爆炸吹开。我们将在第三章"恒星爆发"一节谈论这种爆炸的天体(超新星)。现在,我们只考虑爆炸对星际气体的影响。

超新星会以几千千米每秒的速度喷射一到数个太阳质量的物质。喷射气体的动能约为10^{44}焦。相较之下,太阳在过去46亿年中辐射的能量也不到6×10^{43}焦,可见那个能量有多么巨大。

超新星喷射的第一批气体猛撞到四周原本静止的气体上,将它们挤压、加热、引发振动。这自然会减慢喷射气体的速度,于是它们又被超新星后续喷射的气体撞击。此时第一批气体变慢、受压、受热,因此,超新星周围形成了一层由压缩气体构成的不断膨胀、又厚又热的壳层。壳层内外由速度不连续的气体分隔:在外它撞向静止的气体,在内它使超新星喷射的气体减速。

随着壳层扫过越来越多的星际气体,它也因膨胀而冷却。

如果它在长时间内不受干扰地膨胀，温度就会最终降低到辐射冷却比内外激波加热还快的程度。周围气体的密度越高，就越快达到这种情况，因为壳层的光度正比于密度与质量的乘积。

在第三章的"恒星的形成"一节，我们会看到恒星在星团中形成，所以超新星标志着大质量星团的终点。因此，在前一颗超新星的膨胀壳层的低密度区域内，往往会发生第二次超新星爆发。第二次爆发的壳层会在低密度区域快速膨胀，与第一次爆发的壳层融合。这就形成了**超新星泡**，它会在膨胀过程中继续催生后续的超新星。猎户座有一块恒星生成活动十分活跃的区域，那里会定期发生超新星爆发，它们推动着一块氢原子墙，以超过100千米每秒的速度向我们靠近——这道墙被称为**猎户座斗篷**。

一些超新星推动着高速氢原子壳层远离银河系中平面，并将气体推到银河系引力场中远离盘面的轨道上。其实，银河系的氢原子大约有10%在距离银道面1千秒差距之外。这些气体最终会回到银道面，因此有人说超新星推动着一座**银河喷泉**。

这样的一层冷气体进入轨道后，超新星喷射的气体便有了冲出银河系的通道。这些气体十分炽热，电子很难与离子结合，因此气体由自由带电粒子组成。我们将这种气体称为**等离子体**。在银河系的一生中，这很可能是个重大过程，星系际空间也充满了超新星的喷射物。

如果到处都在不断地形成恒星，那么相邻的超新星泡就会重叠。尽管我们认为银河系平均50年才会发生一次超新星爆发，但超新星泡却在不断重叠，充满了大部分星际空间，将越发致密的星际气体挤压在泡与泡之间狭窄的空间里。理想气体的压强

（P）、密度（n）和温度（T）遵循波义耳定律：$P = $ 常量 $\times nT$。星际气体中热气体施加的压强与冷气体的大致相当，因此，虽然星际气体的温度从20开到2×10^6开不等，但它的温度和密度的乘积nT仍然大致保持为常数。

超新星将电子和离子加速到相对论能量（见第六章的"激波与粒子加速"）。粒子束沿着磁场线在星际空间中流动。每隔一段时间就会有一团相对论性的离子撞击星际气体的原子核，产生γ射线。这种现象在单位空间中的发生率大致正比于气体的密度n，因此，γ射线的强度是确定星际气体密度的有效方式。

至此，我们看到了恒星对星际气体的巨大影响。下面就来看看恒星是怎么运作的吧。

第三章

恒 星

目前,我们探测宇宙最重要的仪器是望远镜,它可以汇聚可见光光子和波长稍长的光子(红外线光子)。在红外波段,整个夜空充满了恒星。我们能分辨出大约十亿颗恒星,它们就像一个个分辨率不高的光点;我们还能分辨一千万亿颗由整个星系的光构成的光斑。这些星系极其遥远,其中恒星众多,我们无法一一分辨。

所以,我们关于宇宙的大部分知识是从恒星研究中一点点获得的。20世纪科学的一大成就便是解释了恒星如何运作,还解释了它们由生到死的生命周期。

恒星的形成

恒星因星际气体云中心密度失控而开始形成,失控的过程已经在第二章讨论过了。气体密度增大好几个数量级,达到一万亿(10^{12})倍后,原子和分子发射的光子便很难从其中逃离,因为它们在致密的气体和尘粒中尚未穿行多远,就会被分子和尘粒散射。给自行车轮胎打气时，打气泵会因为不断压缩空气而变热。同样，引力压缩坍缩的气体云时，如果不将新获得的能量辐射出去，气体也会变热。若是光子无法轻易逃离气体云，压缩做的功就无法及时辐射出去，温度就会逐渐升高。然而，即使气体云中心温度和压强升高，随着越来越多的气体落入气体云中心，引力的挤压力也由此变大。结果就是中心温度和密度持续升高。如果气体云质量足够大，最终温度和密度就会引发**核燃烧**，通过氢核向氦核嬗变，以及氦核向更重的原子核如碳、硅、铁等嬗变释放能量。下面就来谈谈核燃烧。

星际云密度失控后不会只形成一颗恒星，而会形成一群恒星。我们还无法完全解释这个分裂过程，但它是一条重要的经验事实。在不断变形的气体云中，会出现几处密度暴涨的区域，每一处都含有孕育恒星的种子。这些种子积累质量的速率差异很大。最后，少数区域会生成大质量恒星，更多的则是小质量恒星。大恒星的质量可达 80 $M_\odot$，小恒星的质量最低不到 0.01 $M_\odot$。这类恒星大多很暗淡，在生命的任何阶段都无法被我们探测到。

由于原始的气体云是一团拥挤、翻腾的气团，其中的各个种子也在相互运动。一方面，一颗种子会闯入气体向另一颗种子下落的路径上，由此增大自身、削弱其他种子；另一方面，由于种子间的相互运动，一颗种子往往会进入环绕其他种子运动的轨道，从而形成双星。在某些地方，一群相互环绕的种子会在引力束缚下形成星团。不过，我们会在第七章的"缓慢漂移"一节中看到，小型星团并不稳定，往往会演变为双星系统和一颗颗单独的恒星。

随着种子不断积聚质量,它们渐渐演变为恒星,输出核能,对原气体云中低密度部分的影响也越来越大。大质量恒星开始辐射紫外线光子,加热低密度的气体,过程我们已经在第13页①见过了。我们会在第四章中看到,年轻恒星周围环绕着一片做轨道运动的气体,被称为吸积盘。吸积盘会沿着自转轴方向发射气体喷流。喷流猛烈撞击在附近气体上,将它们加热。由此导致的结果是,某些区域的气体云密度失控后,当地大部分气体会在受热后逃离。结果,像是包含 $10^4 \, M_\odot$ 质量的气体云,就只会形成质量约为 $100 \, M_\odot$ 的恒星。在这么低的**恒星生成效率**下,银河系这样的星系在宇宙漫长的历史中以相当平稳的速率生成着恒星,因为这表明星际气体转化为恒星的速率很低。

核聚变

原子由一个微小的、带正电的原子核与一个或多个电子组成,电子在距离原子核大约 10^{-10} 米的轨道上运动。原子核几乎囊括了原子的全部质量,直径却只有大约 10^{-15} 米。两个原子相撞时,电子轨道会变化,使得原子核外的负电荷分布发生变化,并且原子核会受静电力影响,偏离原先的直线轨迹。结果,即便是两个原子迎头相撞,它们的原子核也不太可能撞上,因为在接触之前,它们的速度就会被静电力反转。可以说,原子中的电子为原子核提供了精密的抗冲击封装,保护原子核免受碰撞,除非撞击极为剧烈。

恒星形成过程中,中心的温度不断升高,原子也越飞越快,

---

① 本书所提页码均为英文原书页码,见边码,下文不再一一说明。——编注

碰撞越来越激烈。电子在撞击下脱离原子,因此越来越多的原子成了"裸原子"。不过,相撞的原子核依然很难真正地碰撞,因为它们都是带正电的粒子,会因为静电力而相互排斥。但碰撞终究还是太过猛烈,一些相撞的原子核发生完全接触,核反应由此开始。

核反应能量的规模比驱动我们身体和汽车的化学反应大一百万倍。因此,气体云核心核反应释放的能量一改往日的局面。气体密度很快稳定下来,此时核反应释放能量的速率与热量穿过一层层大质量气体包层向外扩散的速率相同。如果能量释放的速率略低于向外输送的速率,中心压强就会降低,于是核心坍缩,温度和密度升高,核反应速率随之加快。反之,如果核反应释放能量的速率大过向外扩散的速率,中心压强就会升高,于是核心膨胀,温度降低,核反应速率随之降低。因此,释放核能维持了恒星内部的稳定。

几个关键的恒星质量

此时,质量超过 $0.08\ M_\odot$ 的恒星会开始核燃烧。质量超过 $0.08\ M_\odot$ 但低于 $0.5\ M_\odot$ 左右的恒星会将氢燃烧为氦,但无法点燃氦。初始质量在 $0.5\ M_\odot$—$8\ M_\odot$ 范围内的恒星会先燃烧氢,再燃烧氦,但无法点燃碳。初始质量超过 $8\ M_\odot$ 但低于 $50\ M_\odot$ 左右的恒星会将碳燃烧为硅,再将硅燃烧为铁。铁原子核的束缚力最紧密,因此将铁嬗变为任何其他元素都无法获得能量——这些铁原子核构成了核灰。

质量大于 $50\ M_\odot$ 的恒星并不稳定,会在进入硅燃烧阶段前爆炸。我们虽然知道它们不稳定,却不确定这种状态的结果是什么。我们认为,恒星会将大部分质量喷射到星际空间中,只留下一颗黑洞表明它曾经存在。

初始质量小于 0.08 M$_\odot$ 的恒星,温度只够将氘燃烧为氦。氘是氢的同位素,原子核由一对紧紧束缚的质子和中子构成,而不是单单一个质子。氘和氢一样,在大爆炸中创生,在恒星中毁灭。它的丰度只有普通氢原子的十万分之一左右,因此恒星不需要太长时间就能将它耗尽。只燃烧氘的恒星被称为褐矮星。氘耗尽后,星体冷却,变成几乎探测不到的黑矮星。

质量大于 0.08 M$_\odot$ 的恒星,一生的主要阶段是在核心区燃烧氢。由于初始的星际云中有四分之三的原子是氢,所以这一阶段的能源十分充足。而且,氢燃烧时,每个核子(中子或质子)释放的能量比其他核燃料都更多。太阳核心区的氢已经燃烧了46亿年,过程刚刚过半。我们把在核心区燃烧氢的恒星称为**主序星**(图3),理由等讲到第三章的"检验理论"一节自然会明白。

恒星的质量越大,核心区消耗氢的速率就越快,主序期就越短。大质量恒星"挥霍无度":它们诞生时拥有的能源越多,因能源耗尽而"破产"的速度就越快。图3绘制的是不同质量恒星的光度随表面温度变化的图像,定量展现了这个事实。图中,主序阶段的恒星从曲线左端的点移动到标记为"2"的点处。大数字给出了恒星的质量,单位是太阳质量。纵坐标是对数坐标,因此质量为 0.6 M$_\odot$ 和 20 M$_\odot$ 的恒星在主序期的光度相差了上百万倍。于是,质量为 0.6 M$_\odot$ 的恒星能在主序期待上780亿年,差不多是宇宙年龄的六倍,而质量为 20 M$_\odot$ 的恒星处在主序期的时间只有850万年。

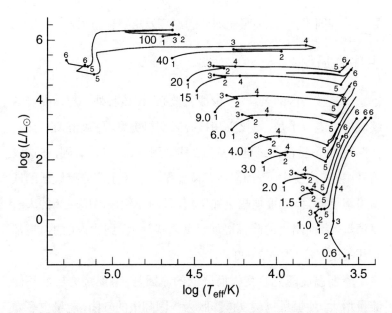

图3 纵坐标的光度以太阳光度为单位,此图绘制了不同初始质量的恒星的光度与表面温度(单位为开尔文)的关系曲线。每条曲线左侧的大数字是恒星质量(以太阳质量$M_\odot$为单位)。恒星处在主序期时,位于曲线左端点附近。恒星开始在核心区将氢转化为氦后,便逐步离开初始位置

  图3还定量表现了另一个关键事实:主序星的表面温度会随质量增大而升高。因此大质量恒星更亮、更热,寿命较短;小质量恒星更暗、更冷,寿命较长。高温加热金属,它会先呈现暗红色,再变为黄色和白色。如果继续升温,金属会发蓝光。因此,炽热的恒星是蓝色的,较冷的恒星是红色的。图3表明所有蓝色恒星的质量都很大,而大质量恒星的寿命都比较短。所以蓝色恒星往往是年轻恒星。

  恒星颜色与温度的关联反映了一项重大的物理现象。**黑体**会吸收打向它的光子并释放辐射。辐射的特征谱只取决于黑体的温度,而与它的材质无关。这样的辐射被称为**黑体辐射**。恒

星大致等同于黑体，按照光球层的温度发出黑体辐射。

主序之后

恒星核心区的氢耗尽、形成氦核心后，核心外一层球壳仍继续燃烧着氢。于是，氦核心和核心区变得更重、更致密，也更热。在图3中，恒星沿曲线迅速从点2移动到点6。可以看到，在这段路径中，光度变大，表面温度却降低了。对于主序光度较低的低质量恒星而言，光度增幅更加剧烈，而对于主序温度较高的大质量恒星，表面温度的降幅更加显著。实际上，到达点6的恒星都有相近的温度，约为2 000开。

表面温度随光度变大而降低的原因是，氢燃烧壳层的核能流量增大，使恒星包层气体膨胀为一团翻滚的气团，能量更多靠对流传递，而不是光子向外扩散。暖气片就是靠对流加热房间：与暖气片接触的热空气上升，再由下沉的冷空气补缺，这些冷空气来自非常冷的窗户和外墙附近。包层转变为一团膨胀的对流团块，扩大了恒星的光球层，这是发出大部分星光的区域。尽管它的温度比之前低，但膨胀的光球层能辐射更高的光度。相比于恒星在核心区的氢耗尽后变成的形态，主序星个头更小，因此它们被称为**矮星**。此后，它们将演变为**红巨星**。

在图3的点6阶段，恒星核心区的氦被点燃。质量大于2 $M_\odot$左右的恒星，氦燃烧比较温和。但在质量更小的恒星中，氦燃烧非常剧烈，我们称之为**氦闪**。氦闪释放的能量会使恒星的包层发生剧烈振荡，将其中相当一部分物质喷射到星际空间。在质量大于2 $M_\odot$的恒星中，第25页描述的调节机制正常发挥作用，使氦燃烧平稳启动。氦燃烧开始后，恒星光度会略为降低，颜色也会稍微变蓝一些。

恒星平稳地燃烧核心区的氦,这是恒星一生中长度仅次于主序阶段的时期。初始质量达到 2.5 $M_\odot$ 的恒星会在这个阶段度过大约 1.3 亿年。质量更大的恒星在这一阶段的时长会随质量变大而迅速缩短,比如质量为 20 $M_\odot$ 的恒星只有 60 万年。

核心区的氦燃烧耗尽后,会形成碳核心。之后,氦燃烧转移到碳核心外的壳层进行。这层壳层之外,氢继续在更靠外的壳层中燃烧,恒星的光度急剧上升。同时,恒星的包层隆起,变得不稳定,不时地向星际空间喷射大量物质。在这一时期,恒星会产生越来越强的风,将初始质量中的大部分物质吹散。随着气体外流,恒星也冷却下来,熔点高的物质形成固体,凝结为尘粒。因此,这些恒星与维多利亚时代的工厂烟囱颇有相近之处。

吹散过程会慢慢失控,因为恒星质量不断减小,对包层中的气体施加的引力也随之减小,使气体越来越难抵抗辐射压。而且,由于覆盖在致密热核外的气体量减少,恒星的光度增加,辐射压也进一步增强——形象地说,就是恒星家里的隔热层消失了。最终,包层底部氢燃烧的区域急剧扩张,包层变成膨胀的气体壳层,包裹着恒星,在恒星此时裸露的核心区照射下十分耀眼。壳层被核心区的光子电离,发出明亮光芒——此时的天体叫作**行星状星云**(图4)。

核心区因核反应终止而逐渐冷却,恒星变成白矮星。当年,钱德拉塞卡在那趟从孟买启程的旅途中,就已经正确勾勒出了它的物理性质(第7页)。

初始质量大于 8 $M_\odot$ 的恒星会点燃碳,将它烧成硅,再将硅烧成铁。由于铁不能继续燃烧,恒星无法再靠核心区获得热能,

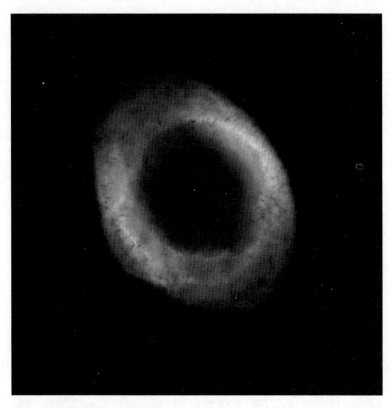

图 4　行星状星云的典型——环状星云

于是转而收缩自身释放引力能。但是，自引力天体在收缩时，中心温度会升高，这对恒星来说是一种致命变化——它会突然爆发成大火球，变成超新星。

恒星的表面

恒星的气体密度由内向外持续减小，幅度先缓后急，但密度不会降至零点。随着密度降低，一般光子被原子散射和吸收前能移动的距离也逐渐增大。而在某个半径处情势突变，这个距离正好与密度减半处的半径不相上下,许多光子可以由此半径逸散到无穷远处。恒星的观测性质很大程度上就取决于这个半径外的气体球壳,也就是**光球层**的物理状况(图5)。

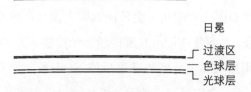

图5 太阳的各个外层。阳光来源于光球层。太阳等离子体的温度在色球层下方最低。温度在过渡区由内向外从大约10 000开飙升到上百万开。炽热的日冕向外延伸极远,在日全食期间很容易观测

不同频率的光子在从恒星中逃逸时,越容易被自由电子散射的光子,逃逸半径就越大。某些光子非常容易被电子散射,因为它们会与普通原子和分子产生共振,结果被困在最大半径处。因此,恒星的亮度会随波长改变而变化,恒星的光谱也会包含各种谱线。谱线的排列蕴含了光球层辐射的密度和温度沿径向变化的信息。于是,天文学家费尽心血整理出大量恒星的高精度光谱。由光谱可以推知恒星的质量、半径、温度和化学组分等信息,但推算的精度则受限于我们计算特定质量、半径等属性的恒星光谱的能力。

日 冕[①]

太阳物质的温度,从太阳中心到光球层顶部逐渐降低。光球层是可见的表面,温度大约是4 500开。出人意料的是,在这

---

① 原文是复数,指一般的星冕。但本节内容只谈及太阳,所以仍译为日冕。

之后温度却突然上升，先缓后急，在只有100千米厚的**过渡区**从大约10 000开飙升到1 000 000开以上（图5）。因为热总是从温度高的物质流向温度低的物质，所以不是太阳加热了日冕，而是日冕在加热太阳。那么是什么加热了日冕呢？难道是外太空？

太阳核心区产生的热量会传向太阳表面。在这个过程的最后一个阶段，大部分热量由对流传递。光球层下方210 000千米处的一小团热气体上升到光球层后停下，不断向外界辐射能量而冷却。最后，它们返回内部重新受热。虽然大体上说，对流是个不断往返的过程，但在光球层中，气体会先横向移动，然后再下沉。因此，对流引起了不稳定的气体环流。

太阳中高度离子化的气体是近乎完美的导体，因为大量自由电子即使在最微弱的电场中也能轻松移动。磁场线会嵌在这样的导电流体里一同运动，因此太阳的气体是磁化的。所以，对流在太阳表面引起的扰乱会影响气体中的场线，使它们不断挤压、拉伸，乱作一团。

磁场线就像有弹性的带子，具有沿场线方向的张力。如果场线拉伸，场强就会增大，张力增强，此时流体对场做功。反过来，如果场线收缩，那么场就对流体做功。

相邻的同方向场线会相互排斥（图6）。如果场的方向正好与表面平行，那么深处的场线就会将靠近外表面的场线向外推开。

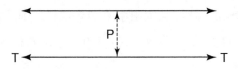

图6　每条磁场线都会受到张力作用，并会排斥方向相同的磁场线

气体不能穿过场线,但可以沿场线运动。如果场线向上隆起,气体就会沿场线离开顶部。气体流失减少了隆起处场线的负担,所以它们抬升得更厉害,使更多气体离开。很快,太阳表面便拱起了一个巨大的磁场环路(图7)。同时,环路两端的致密气体受到对流和太阳自转的双重影响,在太阳表面形成气流;构成环路的场线也会扭作一团,因为方向完全相反的场线会不断靠近。此时,场线会在日冕中**重连**,如图8所示。

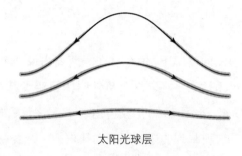

太阳光球层

图7 等离子体从三条向上隆起的场线顶部流走

图8 方向相反的磁场线相互靠近时,相反的部分会相互抵消,释放磁能,然后磁场会变成不同的"重连"样式

场线重连后,磁场中储存的能量被用于加速粒子。这些高能粒子中的大多数会与附近的电子和离子碰撞,加热附近的气体并损失额外获得的能量。因此,在发生重连的区域,气体会变得极为炽热。所以,日冕的灼热温度主要由磁能维持,这些磁能来自光球层之下的湍流层,它们穿过包裹着光球层、由低密度气体构成的厚达2 000千米的色球层进入日冕(图5)。

日冕中的气体大多非常热，无法被太阳的引力场束缚，会逸散出去形成**太阳风**。太阳风会在距离地球大约60 000千米的地方被地球磁场偏转。光球层外因重连而加速到极高能量的电子，不会失去太多能量就进入太阳风中，其中一部分被地球的磁场捕获。这些粒子构成了**范艾伦**辐射带。粒子以接近光速的速度从一个磁极移动到另一个磁极，在靠近北极附近地表时激发空气分子使之发光，这就是北极光的成因。

恒星爆发

在极为偶然的情况下，恒星会在一两周内像一个包含1 000亿颗恒星的星系那么明亮。这类事件叫作**超新星爆发**。在银河系这样的星系中，预计每50年就会出现一次超新星爆发——虽说银河系上一次观测到的爆发还是约翰内斯·开普勒在1604年记录的。我们还发现了1680年和1868年超新星爆发的遗骸，但爆发本身却未见记载。我们在1987年2月观测到了大麦哲伦星云中的超新星SN1987A。大麦哲伦星云是个小型星系，目前正以极近的距离穿过银河系，它最终会被银河系吞并。那场爆发是迄今为止我们观测超新星条件最好的一次。

超新星可以远距离观测，因此是重要的宇宙工具。于是，近年来，观测资源主要用在了探测大量超新星方面。

结果表明，超新星可以通过两种完全不同的机制产生。

核坍缩超新星

恒星将碳燃烧为硅时，核心区内部极度致密，对抗引力的压力主要由电子提供。电子遵循海森堡原理和泡利原理（第7

页),相比同温度、低密度的情况,此时电子的无规则运动要快得多。因此,电子具有极高的动能,从能量方面看更容易进入原子核中,降低原子核的电荷量,使之转变为周期表上前一位的原子。每一次捕获都会降低电子数量,从而降低对抗引力的压力。

随着核心区收缩,它的温度逐渐升高,附近黑体辐射的光子平均能量也随之升高(第28页)。最后,辐射中包含大量高能光子,足以将原子核轰成碎片(这一过程被称为**光致离解**)。核心区的光子气体为对抗引力发挥了重要作用,而光致离解则吸收了光子气体的能量,降低了光压。

因此,恒星开始走"下坡路":收缩提升温度,升温使更多电子被捕获,同时引起更多光致离解,导致进一步的收缩。核心区在几毫秒内处于自由落体状态,灾难无可避免。

随着中心密度升高,恒星一生中默默组装起来的各种原子核也分崩离析,其中大部分碎片是中子。此时,中子开始发挥先前电子发挥的作用:它们也要遵循海森堡和泡利原理,速度远大于同温度、低密度的情况,提供了大部分压力。于是,在某个范围内,核心区的压力随密度急剧增大,突然间不再收缩,甚至开始"反弹"。

恒星的大部分质量在这个靠压力支撑的核心区之外,并在急速向中心坠落。结果就是形成激波(见第六章的"激波与粒子加速")——向内坠落的物质瞬间停止,剧烈升温。

此时温度和密度都非常高,电子、中子和质子组成的等离子体内部碰撞不断,产生了大量中微子。因为中微子的反应截面非常小(第6页),哪怕在恒星极其致密的中心,它们也能在两次碰撞间移动很远的距离。因此,核心区因为发射中微子而变得非常明亮——它也发射光子,但光子向外扩散的速度很慢,因此中微子带走的能量要高得多。在这一阶段,大量中微子贯穿恒星包层逸出,而包层仍在向几乎压缩成一个点的核心区下落。一小部分中微子与下落的原子核碰撞,将能量和动量传递过去。传递的能量足以逆转包层大部分物质的向下运动,将它们向外吹散,形成巨大的火球。

核心区富含中子,会喷射强烈的中子流。包层物质散入星际空间前,会先受到中子流冲击。几乎所有中子都会被包层中的原子核吸收,使它们转变为更重的核子。一般而言,通过吸收中子形成的原子核会具有很高的放射性,随后迅速衰变为其他核子。这个过程常常会发射一个电子,而且几乎总会发射一个光子。因此,放射性衰变成为扩散的包层中的主要热源。有些核子会逐步吸收多个中子,再经历多次放射性衰变。元素周期表中所有排在铁之后的元素都以这种方式形成,也就是说,溴、银、金、碘、铅、铀都产生于超新星爆发。

随着火球扩张,光球层也一并膨胀,光度因此增大。以SN1987A为例,它的光学光度在核心区内爆后三个月达到峰值——我们之所以知道发生过内爆,是因为探测到了核心区反弹产生的中微子风暴。(目前为止,在所有超新星中,我们只在SN1987A那里探测到了中微子。)这一阶段的光谱表明,包层物质以几千千米每秒的速度飞离核心区。按这个速度,每太阳质量的喷射物将具有2×10^{43}焦动能。因此,大约$5\,M_\odot$的喷射物包含了10^{44}焦左右的能量,这只是核心区坍缩释放的引力能的1%左右。以上便是超新星的一个典型特征:我们见到的壮观的爆发,以及爆发对星际介质的巨大影响,不过是实际释放的能量中的一小部分引发的。99%的能量被中微子带走了,而它们不会,也永远不会与任何物质发生相互作用。

最终,膨胀的包层变得非常稀薄,光子在其中大部分区域不会再受到长时间的束缚。因此,超新星的光度会在几周内逐渐下降。等到膨胀得更稀薄时,外围气体中的动力学碰撞变得重要起来。其实,这里的气体密度可能异常地高,因为恒星在内爆成超新星之前,便以恒星风的形式迅速吹散物质。恒星爆发的冲击波急速冲向之前的风,产生激波,使它们变热。在这一阶段,可以在无线电波段探测到显著的发射信号。

与此同时,核心区也发生着许多事情。由前文可知,坍缩的核心区充满了中子,而且中子产生的压力使核心区反弹,引发中微子暴,将大部分包层物质吹散出去。反弹过后,物质再度落回稳定的核心区,中子星由此成形:它是一个巨大无比、充满了中子的原子核。其中的中子以刚刚达到的相对论速度四处乱撞,提供了对抗引力的压力,和白矮星中电子对抗引力的方式相同。中子星的质量随物质不断落入而增大。质量越大,半径就越小(白矮星也如此),引力场越强。中子为对抗引力,运动得越来越快。值得注意的是,广义相对论认为,压力本身也是引力源,所以对抗引力的压力越强,引力也越强。如果中子星的质量大于某个临界值$M_{临界}$,引力就会超越中子星的压力,整个星体将坍缩成黑洞。

$M_{临界}$的精确值仍有争议,因为它取决于物质在核子级密度下的行为。虽然我们自认为知道要解什么样的**量子色动力学**(QCD)方程,但我们很难从第一原理确定密度与压力的关系。此外,我们也没有通过实验验证相关结构的好办法——地球上最重的核子只包含大约240个质子和中子,产生的引力场微乎其微。不过，专家相信 $M_{临界}$ 在 1.4 $M_\odot$ 到 3 $M_\odot$ 之间，因此某些核坍缩超新星会留下中子星，某些则会形成黑洞。中国人在1054年记录的超新星就留下了一颗中子星（即**蟹状星云脉冲星**），它已经得到了充分的研究。我们对SN1987A的看法是，它应当形成中子星，但多项更严密的研究尚未确定它留下了什么星体。

暴燃超新星

由前文可知，初始质量小于 8 $M_\odot$ 的恒星不能点燃碳原子，并会失去包层，只留下碳和氧构成的核心，渐渐冷却成白矮星。如果恒星没有伴星，那么它的一生便就此终结。不过，大部分恒星拥有伴星，它们的未来会更加精彩。如果伴星的初始质量更低，它的演化时间也会更长。于是，伴星会在同伴变成白矮星后膨胀，并以极高的速率吹出物质。如果两颗星体相距不远，伴星吹出的大部分物质就会被白矮星的引力场捕获，形成吸积盘。这种吸积盘可以通过X射线辐射来研究（见第四章的"时域天文学"）。气体顺着吸积盘内沿落入白矮星，使白矮星质量增大。

随着质量增大，白矮星的半径减小，引力场变得更强。海森堡和泡利原理使大部分能量高的电子加速。最终，其中一些核子会快到足以触发碳合成为硅的反应。这个过程释放能量，加热星体，使之发生更多的核反应。

如果星体内的压力主要来自核子的热运动，那么核反应产生的热就会使星体膨胀，温度降低。这两个结果都会减慢核反应，使系统趋于稳定。但在白矮星中，压力的主要来源并非核子，而是电子。因此，它的密度不会因为核子受热而减小，核反应速率会急剧增大，直至失控。形容这种核反应速率失控的术语是**暴燃**。这是一种慢速爆炸，高温以及反应速率的波前会以低于音速的速度在介质中传递。

不到一秒的时间内，就有大约一个太阳质量的碳和氧被一路燃烧成铁和镍。瞬间释放的能量会产生极大的压力，足以盖过引力，将星体炸碎，原先的位置则没留下任何靠引力束缚的物体。飞散的物质中，有很大一部分包含了镍的高放射性同位素 ^{56}Ni，它的半衰期是6.1天。^{56}Ni 会衰变为铁，释放的 γ 射线将飞散的物质加热，使它们发光。我们就是由此观测到暴燃超新星。我们根据超新星的光谱，按经验分类将它们称为 **Ia 型超新星**。

Ia 型超新星对天文学的重要性体现在两个不同的方面。其一，它证明了可以通过亮度衰减的速率估算 Ia 型超新星的光度。因此，我们可以将超新星用作**标准烛光**。已知光度的物体，可以通过视亮度推算它们与我们的距离。其二，Ia 型超新星是铁的主要来源，因为原本的白矮星几乎全部转化成了铁。核坍缩的超新星则不同，它们会产生各种重元素，其中**阿尔法元素**（包括碳、硅、镁、钙等）尤为丰富。因此，只要测量恒星中的铁相对于阿尔法元素的丰度，就可以判断 Ia 型超新星和核坍缩超新星对恒星物质增丰的重要程度。我们将在第七章的"化学演化"一节看到这个判断的作用。

检验理论

恒星演化理论需要输入大量原子物理以及核物理的信息，还需要大量数值计算和一些有关湍流如何混合的假设。这些都能保证准确无误吗？我们认为，它们基本上可靠，因为现在已经可以从多方面比较预测结果与实际的观测结果。

球状星团

在恒星演化理论的形成阶段,球状星团是经典的检验对象。银河系中大约有150个近乎球状且密集的星团。以NGC 7006(图9)为例,距离它中心大约10秒差距的范围内有数万颗恒星。相比之下,太阳附近10秒差距范围内只有不到100颗恒星。球状星团的这个特点使它成为恒星演化理论的理想检验对象。结论可以很好地简化为:其中的恒星只有质量差异,它们的年龄、

图9　球状星团 NGC 7006

化学组分，以及与我们的距离都相同。此外，恒星的化学组分可以通过光谱推测。对任何年龄与距离的星团，理论都预测恒星将位于**颜色-星等图**上一条叫作**等龄线**的曲线上，这条线按恒星亮度与颜色的关系绘制，如图10所示。由于历史原因，蓝色（表示表面温度高）放在了图的左侧。纵轴的亮度通常采用对数坐

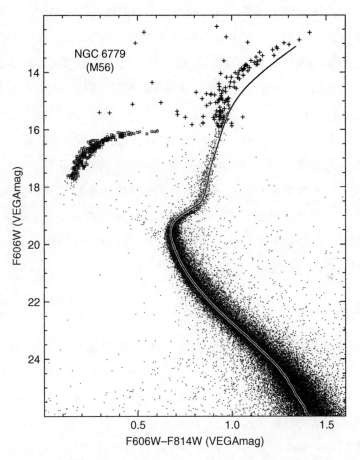

图10 球状星团NGC 6779中恒星亮度与颜色的关系图。蓝色的恒星位于左侧，明亮的恒星位于顶部。图中的曲线是理论等龄线（见正文）。从右下方一直到拐点的点构成了主序。主序恒星在核心区燃烧氢

标，因此改变与星团的距离会使恒星构成的等龄线上下移动（到对应距离的位置），但不会改变它的整体形状。改变星团的年龄则会改变等龄线的形状，这是可以计算的。找到最符合观测到的恒星分布的等龄线，再找到最匹配的纵坐标位置，就可以确定星团的年龄与距离。

由图10可以看到，用这种方式得到的理论曲线与观测结果匹配得非常好。不过其中也确实存在些许差异，天文学家仍在努力完善模型的数据和假设，以减少差别。理论结果与许多星团的数据吻合，因此对于理论基础是否牢靠的问题，我们基本上没有任何疑虑。

我们也从中得到了一个有趣的结果：银河系中的球状星团都非常古老——具体的数值一度与宇宙年龄产生矛盾。之后，我们对宇宙膨胀和恒星演化都有了更准确的理解，二者的年龄也不再冲突。"金属"元素（指元素周期表中排在氦之后的元素）丰度最低的星团往往最古老，它们的年龄大约是120亿年，没有超过宇宙的年龄（约为138亿年）。

太阳中微子

在太阳中，每两个质子聚变为氘，并随即聚变为氦时，都会产生一个中微子逃离太阳。因此，太阳除了辐射光子外，还在辐射中微子。1960年代，雷伊·戴维斯着手准备探测太阳中微子，他认为这是对恒星演化理论的重要检验。因为中微子直接来自太阳的产能核心区，而光子来自光球层，所以我们可以探测到一片完全不同的区域。

戴维斯实验的做法是，将数吨干洗液（四氯化碳CCl_4）灌入矿井中——只有在矿井内才能隔绝宇宙射线,否则它们会产生繁杂的背景信号。四氯化碳可以便捷地储存 ^{37}Cl 原子核,它被中微子击中后会变成氩。之后便将氩提取出来测定质量。戴维斯经过多年艰苦工作,探测的太阳中微子流量只有预期的三分之一。科学界对戴维斯的失败并不意外:有人怀疑这个实验并不能那么有效地检测氩,还有人怀疑预期的中微子流量不准确。

这些疑虑之所以让人困扰,是因为实验只对太阳产生的一小部分中微子比较灵敏:质子聚变产生的中微子,能量不足以将氯嬗变为氩;而戴维斯希望探测的能量更高的中微子来自另一类反应,这类反应对温度极为敏感,产生的太阳能量占比不大。从太阳核心区带出热量的速率即便只有轻微变化,也会大幅降低这种高能中微子的流量。因此,各位专家重新检验了他们的太阳模型,但还是无法降低高能中微子的流量,使它满足戴维斯的实验结果。

另一个问题是,戴维斯的实验只对一种中微子比较灵敏,而中微子共分三种,分别与电子、μ子和τ子相关,戴维斯的实验只能探测电子中微子。这本不应该成为问题,因为核反应产生的应当是电子中微子。但穿过戴维斯实验室的中微子中,是否有三分之二因为某种方式变成了无法探测的μ子中微子和τ子中微子呢?

1930年代中期之后,科学家又建造了两座大型中微子探测器,分别用纯水(H_2O)和重水(D_2O)作为探测介质。它们的最大优势是对三种中微子都非常灵敏。其中日本的**神冈探测器**探测到了SN1987A的中微子暴,加拿大的**萨德伯里中微子天文台**则证明,如果计入三种中微子,那么太阳的中微子流量就符合最初的模型预期。萨德伯里中微子天文台的结果证实了一个观点：中微子在离开太阳后，会从电子中微子转变为另外两种中微子，等它们到达地球时，三种中微子的数量大体相当。因此，天体物理学对太阳的解释从一开始就是对的，戴维斯实验的问题存在于粒子物理层面。

中微子会在不同类型之间振荡的观点，原本只是用来解释太阳中微子实验的结果。不过，我们利用核反应产生的中微子束，对其中的过程做了颇为细致的研究。这个现象非常重要，因为它表明中微子的静止质量不是零。许多实验将电子中微子的静止质量限定在小于2电子伏的小范围内，但中微子振荡确定了中微子具有非零的静止质量。

恒星地震学

恒星和铃铛一样，有天然的振动频率。每种频率都与特定的振动**模式**或者说类型有关。对流产生的气流泡激发了恒星振荡，通过测量恒星振动的频谱，可以获得大量有关恒星结构的信息。因此，大约在1985年之后，人们启动了一些大型观测项目，先是监测太阳，进而再监测附近较为明亮的恒星，对它们的振动频谱做了测绘。

恒星的振动模式有两种基本类型。容易理解的是p**模式**，它与管风琴管的振动模式相似：琴管或恒星中的气体交替压缩、扩张，形成驻波。从天体物理的角度看，p模式不如另一类模式——g**模式**有趣。g模式有点像海波：低密度的流体（空气）在高密度流体（水）的上方，高密度流体表面的波在平衡面上下振动，可以在两种流体的交界面传播。对于海洋表面的波，上下流体的密度不连续；但海洋内部因为盐分不同而存在密度梯度，咸水的密度比淡水更大。因此在平衡态下，更咸的水在更淡的水下方，破坏等盐度面的波可以跨洋传播。

在恒星中，低**熵**气体位于高熵气体下方。熵可以度量流体的热混乱程度。向流体中传入热量，会使熵升高；从中抽出热量，会使熵降低。它与温度不同：打气筒或内燃机气缸中的空气被压缩时，气体的温度会升高，但熵保持不变。在恒星等熵面振动的波，传播方式与海洋中在等盐度面传播的波相同。g模式就是这种类型的驻波。

石油公司勘探石油的方法是，先利用爆炸引发地震波，再用远程传感器探测。之后，计算机根据地震波在地球内从震源传播到传感器的方式算出当地岩石的密度和弹性等性质。恒星振荡模式的频谱同样能灵敏地反映恒星内不同层次气体的密度与旋转速度。因此，只要有合适的软件，我们就可以大致确定恒星内部的密度和旋转速度。这些值可以与理论模型的预测值比较。模型最难确定的是恒星的年龄，在实践中必须将模型与观测数据比对才能估算。恒星简正模的振动谱很大程度上限制了年龄范围，因为恒星中心的致密程度会随年龄增长而变大：核心区收缩，变得更热、更致密，包层则在膨胀。这种变化会改变振动频谱。

日震学的结果表明，用来自核物理和原子物理等领域获得的广泛数据建立的太阳模型非常有效，但并不完美。模型预测与日震观测结果有细微差异，这可能是原子数据或恒星模型的局限造成的。但它也可能暗示，恒星向外传输能量的背后可能存在全新的物理现象："暗物质"粒子（见第七章）可能困在了恒星内部，因为它们很难被其他粒子散射，所以在热传播的过程中占据了不成比例的份额。

双 星

至少半数恒星组成了双星系统。双星系统本身也是恒星演化理论中重要的研究问题，因为质量更大、演化更快的恒星演变为红巨星后，它的伴星很容易从膨胀的包层中夺取气体。这种"盗窃"行为使两者都偏离了原本单颗恒星的演化路线，因为质量是决定恒星演化的关键，而此时两颗恒星的质量都成了时间的函数。

物质落向质量较小、密度更高的恒星上时，会辐射能量（见第四章）。其中一部分会加热大质量恒星的外层气体，使它们热到脱离双星系统，形成恒星风。

不论是一颗恒星向另一颗恒星传递物质，还是以风的形式损失气体，都会改变双星的轨道，使两者更加靠近。一旦双星靠近，物质传递和喷射速率就会加快，导致过程失控，最终二者合并。实际上，即使双星的轨道尚未变化到那种程度，大质量恒星膨胀的包层或许就已经将小质量恒星裹入其中了。

一旦小质量恒星进入大质量恒星内部，它的轨道运动就会受阻。于是包层变热，小质量恒星螺旋向内移动。一段时间之后，系统变成一颗恒星，但它的核心区与包层都不可能是单颗恒星演化的结果。

简而言之，紧密的双星就像是一个复杂的潘多拉魔盒，尝试解释盒中的内容，是当今研究的一块热门领域。

第四章

吸 积

洗净碗碟，给水槽放水，可以看到水在流进下水道前，会在排水口附近形成涡旋，围出一段空气柱。水流之所以旋转，是因为流入下水道时，水必须保持角动量守恒。每单位质量的角动量由公式

$$L = rv_t$$

给出，其中 r 是水流到环绕点（下水道管道的中心）的距离，v_t 是水流速度垂直于指向中心点方向的分量。水流向内流动时，r 不断减小，v_t 不断增加，使 L 保持恒定。龙卷风（旋风）的物理原理也一样，但效果夸张得多：气流向龙卷风内部流动，取代温暖、潮湿的上升气流时，会围绕龙卷风中心旋转得越来越快，乃至于能够掀起屋顶，吹翻汽车，造成严重破坏。

角动量守恒定律是银河系形成盘状结构的重要因素，也是银河系中的太阳形成太阳系的关键。我们将在本章看到，这条定律在宇宙中许多最奇特、最明亮的天体身上也发挥了很大的

作用。

吸积盘

在这类系统中,引力将气体吸向某个中心,这个中心可以是星系中心,也可以是恒星或黑洞的中心。总之,这些天体在**吸积**气体,而角动量守恒使吸积的气体在向内运动时围绕吸积天体旋转。如果气体温度低,压强不足以有效对抗向内的引力,那么这些旋转的气体就会形成**吸积盘**。吸积盘中,各半径处的气体大体沿着圆周轨道环绕吸积中心运动。

吸积盘的基本动力学

考察吸积盘结构的一条思路是,设想它由大量固体圆环构成,每个环都在自转,仿佛其中的粒子都在围绕中心物质做轨道运动(图11)。在距离太阳或黑洞等致密物质 r 处的圆周轨道上,粒子的速度正比于 $1/\sqrt{r}$,因此越靠近中心,速度就越快。由此可知,吸积盘内部存在**剪切力**:圆环不会与外部相邻的环同步旋转,而是会在流体内的摩擦力或者说**黏性**作用下,沿旋转方向对外部相邻的环施加**扭矩**,使它旋转得更快。

扭矩与角动量的关系相当于力与动量的关系,表示角动量的变化率。由牛顿运动定律可知,力等于动量的变化率。同样,角动量的变化率等于物体受到的扭矩。因此,较小的环对较大的环产生扭矩,说明角动量在由内向外沿着吸积盘传递。等到吸积盘状态稳定时,因黏性向外传递的角动量,与向内传递的角动量达到平衡,因为气体螺旋进入吸积盘时,自身也携带着角动量。

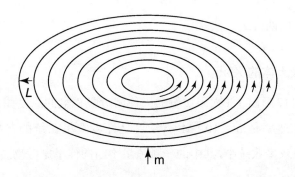

图11 设想吸积盘由许多固体环组成,它们以不同的速率绕公共轴旋转,其中弯箭头的长度与该位置的速率成正比。圆环旋转一周所需的时间由内向外逐渐增加。质量沿吸积盘向内流动,角动量L则向外流动

气体穿过某道圆环外沿时的能量,比它从这个圆环内沿离开时更多,因为引力势能减少量是旋转动能增加量的两倍。所以,每份穿过圆环的气体都会在圆环内留下一部分能量。而且,下一道环还会沿旋转方向对当前圆环施加扭矩,对它做功。它的功率比当前圆环对上一道环做功的功率更高。因此,当前圆环也会从由黏性驱动、穿过吸积盘的角动量流中获得能量。由于向内的物质流和向外的角动量流都提供了能量,因此环内物质开始发光。这就是带有吸积盘的系统都非常明亮的原因。

为较好地估算结果,假设每道环都像黑体一样辐射(第28页),因此吸积盘的辐射谱可以由温度随半径变化的函数$T(r)$计算。平衡时,能量聚集的速率与辐射带走能量的速率相等,由此可以计算平均半径为r的圆环的温度。如果吸积天体以$\dot{m}$的速率获得质量,那么吸积盘的温度由

$$T(r) = \left(\frac{GM\dot{m}}{2\pi r^3 \sigma}\right)^{1/4} \tag{4.1}$$

给出。温度以负四分之三次幂随半径增大而降低,且正比于吸

积速率的四分之一次幂。

恒星级天体的吸积

图12画出了1太阳质量的天体附近吸积盘的温度图像。它的吸积速率是 $\dot{m} = 10^{-8}\,M_\odot/\mathrm{yr}$，这是许多双星系统的典型推测值。在双星系统中，其中的一颗恒星不断向伴星抛撒物质（见第三章的"双星"）。

横轴和纵轴都取对数坐标，范围从恒星级黑洞的半径（以 $R_\mathrm{s} \simeq 3\,\mathrm{km}$ 为典型）一直到冥王星轨道半径（第72页），温度则从 R_s 的1亿开（比太阳核心区温度高好几个数量级）经太阳半径处的太阳表面温度 $T_\odot$，一直下降到地球轨道半径的100开。如果吸积天体是黑洞，那么图中的半径范围大体上都会与它物理相关；如果吸积天体是太阳质量的白矮星，那就只有标记了 R_wd 的

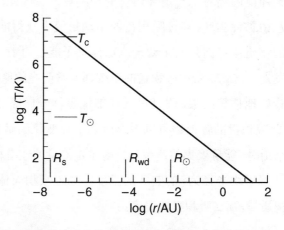

图12　1太阳质量的致密天体的吸积盘在各半径（r）处的温度。天文单位（AU）是地球公转轨道的平均半径（第8页）。假定吸积速率为 $10^{-8}\,M_\odot\cdot\mathrm{yr}^{-1}$。图中还标出了太阳中心和表面的温度 T_c 及 $T_\odot$，以及1太阳质量的黑洞、典型的白矮星和太阳表面的半径 R_s、R_wd 和 $R_\odot$。

那根线右侧才有明显的物理影响,R_{wd} 对应温度是 200 000 开;如果吸积天体是类似太阳的恒星,那么相关区域只有标记 $R_\odot$ 的那根线右侧,对应温度 $T = 4\ 700 \simeq T_\odot$。相应温度下,吸积盘在 R_s 外侧附近主要辐射 X 射线,在 R_{wd} 外侧附近辐射 X 射线和紫外线,在 $R_\odot$ 外侧附近主要辐射可见光。

图 13 画出了吸积盘半径 r 处(横坐标)辐射的光度。可以看到,在地球轨道处,吸积盘辐射的光度只有太阳光度的千分之几;在太阳半径处大约是 $L_\odot/2$,在白矮星半径处大约是 $60\ L_\odot$,而到了黑洞半径处约为 $100\ 000\ L_\odot$。结合图 12 得出的结论,可知吸积速率为 $10^{-8}\ M_\odot\ yr^{-1}$ 时,1 太阳质量的黑洞是极为明亮的硬 X 射线源,同样的白矮星是明亮的软 X 射线源,主序星则会因为吸积盘大幅提升自身的光度。相比上述情况,地球轨道外的吸积盘只能发出红外线,而且在吸积盘和吸积天体内部极高的光度反衬下很难探测。不过,这部分吸积盘非常重要,因为那里

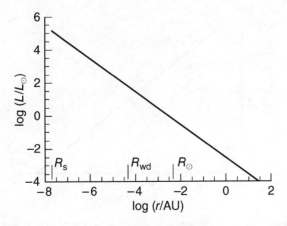

图 13 1 太阳质量的天体的吸积盘在半径 r 处辐射的光度。半径的单位是地球公转轨道半径,光度单位是太阳光度。吸积速率假定为 $10^{-8}\ M_\odot\ yr^{-1}$

是行星形成的场所（见第五章的"行星的诞生"）。

类星体

图12和图13给出了恒星级黑洞的几个特征值。至于星系中心的黑洞，则需要考虑完全不同的尺度。这些黑洞的质量在$10^6\,M_\odot$到$10^{10}\,M_\odot$之间，而且它们为类星体提供能源，吸积速率必须是$1\,M_\odot\,yr^{-1}$数量级。因此，它们的质量是恒星级黑洞的10^8倍左右，吸积速率快10^8倍。又因为黑洞的特征半径R_s正比于质量，所以这些黑洞的半径也大10^8倍左右。由公式（4.1）可知，在黑洞特征半径R_s的几倍处，与1太阳质量黑洞相比，特大质量黑洞吸积盘的温度会低两个数量级，光度会高八个数量级。图14和图15反映了这一事实。

这种黑洞的特征半径稍大于地球轨道半径，对应的温度只有大约100 000开，因此辐射的主要是软X射线和紫外线。这种

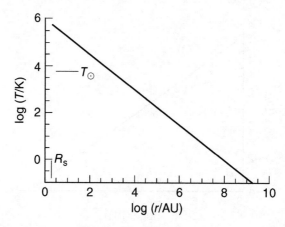

图14 质量为$10^8\,M_\odot$的黑洞外吸积盘的温度。这类黑洞常见于星系中心，吸积速率是$1\,M_\odot/yr$。R_s标出了黑洞半径。图中还标出了太阳表面温度$T_\odot$，以作比较

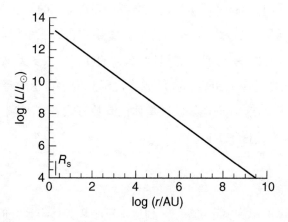

图15 以1 M☉/yr的速率吸积, 质量为10^8 M☉的黑洞的吸积盘在各半径(r)处辐射的光度

系统的光度大得惊人。距离吸积盘中心600天文单位(比冥王星轨道半径大10倍)外的部分,光度与整个银河系1 000亿颗恒星不相上下。此半径到黑洞表面之间那部分的光度还要大100多倍。

因为光度极大,而且主要呈现为很容易观测的紫外线和可见光,所以吸积中的特大质量黑洞在宇宙中随处可见。天文学家最初获得类星体光谱时,并没有做出正确的解释,因为他们没料到谱线向红光方向有很大的频移。移动量用**红移**z表示,由光从类星体发出时以及在地球上观测时的波长关系确定:

$$\lambda_{观测} = (1+z)\lambda_{发射}$$

因此,$z = 0$表示没有红移,$z = 1$表示观测到的谱线波长是它们出射时的两倍。1963年,马丁·施密特提出了一个石破天惊的结论,他证明天体3C-273具有红移$z = 0.158$。继施密特的论文之后,科学家又证明其他光谱源还存在更大的 z 值。对于这些红移现象,最传统的解释是,它们都由宇宙膨胀引起。然而,当时并没有发现任何星系具有这么大的红移。很多天文学家怀疑,这些光谱源的距离和亮度或许达不到用宇宙论解释红移时的要求。现在,我们已知**准星体**已有数万颗之多,其中一些的红移 z 甚至超过了 7!

旅途的终点

气体沿螺旋穿过吸积盘,到达临近吸积天体半径处时,就必须从吸积盘进入星体了。气体进入的方式会对身后吸积盘的结构产生影响。我们以吸积白矮星为例,分析其中的要点。在吸积盘内沿,气体相当于在星体表面做掠地圆周运动。维持轨道运动的动能是每单位质量 $E_c = \frac{1}{2}GM/r$,数值上相当于气体从无穷远处螺旋移动到 r 处损失的所有能量。因此这个能量非常大,而且如果气体坠入星体,它的能量会瞬间转化为热。由于气体最终都会坠入星体,因此星体和吸积盘之间形成了一层极热的气体薄层。这块**边界层**自然增大了整个系统在最短波段的辐射强度。

前文提到,吸积盘中每个圆环辐射的能量中,有四分之三来源于内侧圆环对它做功与它对外侧圆环做功的差值。但最内侧的环是个例外,因为它对外侧的环和内侧的边界层都做功——边界层因为星体的迟滞而只能缓慢旋转。因此,尽管边界层异常明亮,吸积盘最内层的环却比普通情况下还要暗淡许多。

这种固态星体缓慢自转、吸积盘延伸到星体表面的情况并不是普遍现象。一些吸积天体(尤其是中子星和白矮星)蕴含着强大的磁场。磁场线由星体发出，在附近绕行到另一处地点重新进入星体，就像地球的磁场线从北极出发，再在南极回到地球一样。如果一颗带有磁场的星体在自转，它的场线就会在星体周围扭转；如果磁场够强，它甚至可以对距离星体很远的气体施加显著的作用力，使气体能够以慢于一般情况下的速度沿轨道旋转。这种情况下，气体会在某个临界半径r_B处离开吸积盘，进入途经此处的场线（相比轨道气体而言，场线的移动速度慢得多），并沿场线向星体的磁极移动（图16）。这种情况便没有又薄又热的边界层，只有两个热极点位于星体的两极附近。气流沿着场线在那里一头撞向星体。

气流沿场线撞向星体的同时，也将自己的角动量通过磁场传给了星体，由此加快了星体的自转速度。随着自转加快，r_B不

图16　气体被武仙DQ型星带出吸积盘

断变小，吸积盘向内扩张。逻辑上看，这个过程的终点是 r_B 缩小到星体半径，届时星体自转极快，赤道上的物质相当于在做轨道运动。此时，星体可以说在以**裂解**状态自转。实际上，这种极端情况不可能发生，但吸积盘确实可以让中子星的转速接近裂解速度。

黑洞外部的吸积盘则以另一种方式终结。爱因斯坦的广义相对论认为，环绕不自转黑洞的圆周运动在半径 $6R_s$ 处具有最大的角速度。因此，在这个半径处的环会对两侧的环都做功，所以比牛顿力学中对应的情况更暗淡。在 $3R_s$ 处，轨道运动变得不稳定，粒子可以不损失任何能量就螺旋进入黑洞：它们的能量被黑洞吞没了。因此，黑洞外围没有热边界层，吸积盘的最内沿也比用牛顿力学计算的结果更暗。黑洞自旋的方式与白矮星相同。

磁场的影响

在第三章的"日冕"一节中，我们看到磁场线会被导电流体牵引并不断地与流体交换能量，因为场线会产生张力，拉动流体。由于吸积盘内部存在转速差异，初始位置相近的两个点会很快分开，除非它们与吸积盘旋转中心的轴线距离正好完全相同（图17）。假设两点在同一条场线上，随后场线会拉伸，张力增大。此时，由图17可以看到，场线拉伸后，张力方向与最初在较小半径处质元的旋转方向相背，同时又拉着靠外的质元沿旋转方向运动。实际上，磁场会将靠里质元的角动量转移给靠外的质元。这与我们预期中黏性的作用效果相同！而且，因速度差而拉伸的场线还会使磁场增强。所以，不论初始的磁场有多弱，它都能不断增大，直至强度足以影响系统的动力学，遏制场

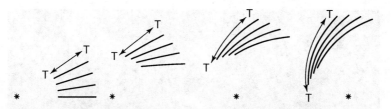

图17 磁场线因吸积盘的转速差而拉伸的过程示意图。恒星位于每张图的底部,最左边的是第一张。吸积盘按逆时针方向旋转。在第一张图中,磁场线长度较短,沿径向分布。而在最右边的最后一张图中,磁场线变长了许多,改为沿切线方向分布

线无限拉长。上面描述的是**磁转动不稳定性**的基本内容,它是史蒂芬·拜尔巴斯和约翰·霍利在1991年为解释吸积盘黏性的来源而提出的理论。

喷 流

磁转动不稳定性不仅解决了长期以来困扰人们的难题——为什么吸积盘的黏性会像观测结果需要的那么大,更重要的是,它提供了一条线索,为吸积盘最显著的一个特征指明了起源的方向,这个特征就是喷流。只要我们有理由认为某个致密天体在吸积物质,那么不论它在何处,观测到的数据主要都是**外向流**,而不是内向流。图18中的赫比格-哈罗天体就是一个典型的例子。它的核心是一颗正在成形的恒星,但它最醒目的特征是一对狭长的喷流,其中极冷的气体以大约200千米每秒的速度喷射而出,比气体中的音速还要快。地球上的宇航工程师只能在梦里想象这种事情。

SS433是另一个呈现喷流形成的著名案例。它由一颗质量远超太阳的恒星与一颗黑洞组成,恒星绕黑洞做轨道运动。恒

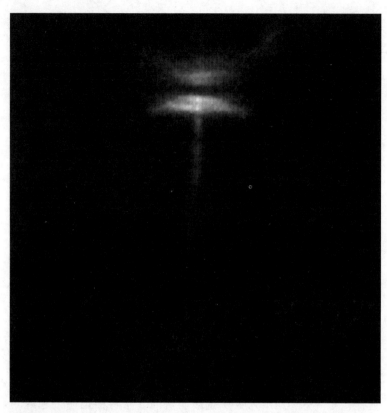

图18 赫比格-哈罗天体HH30。喷流自一颗原恒星的吸积盘两侧喷出。吸积盘本身很暗,呈漏斗状,在散射光的明亮背景下才反衬出它的剪影

星不断向黑洞输送气体,气体则通过吸积盘螺旋进入黑洞。两道喷流携带物质,沿吸积盘转轴以超过四分之一光速的速度($0.26\,c$,其中 c = 300 000 km/s,是光速)射出,而气体的温度够低,氢原子还能存在,并向外辐射氢原子的特征谱。同样,喷射气体的速度比音速快得多。

SS433吸积盘的自转轴还在绕一根轴进动,这根轴与天空平面的夹角不足11°(图19)。在转动中心,喷流总沿着当前的自

转轴喷出。气体离开黑洞后沿直线运动,但它们身后的盘面却在进动,将后续气体沿新的自转轴喷出。所以喷流整体呈锥体螺旋状分布,锥角是40°(图20)。

最后一个例子的尺度更大:射电星系天鹅座A,如图21所示。其中,两条极细的等离子体喷流从一个巨型椭圆星系中心射出,距离黑洞0.77兆秒差距,以惊人的力量撞向星系际介质。在这个过程中,电子被加速到极高的能量(见第六章的"激波与

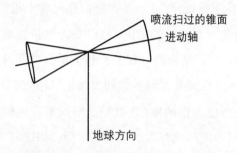

图19　SS433的几何形状。两道相反的喷流扫过锥面。锥体的半开角是20°,轴线与指向地球的连线夹角为80°

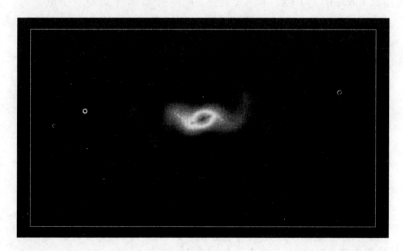

图20　双星SS433螺旋状喷流的射电图像

粒子加速"）。高能电子发射的射电波，显现出图21中整个系统的样貌。

这三个例子各有不同的特征。第一个例子是年轻恒星吸积，后两个是黑洞吸积。前两个是恒星质量级天体引起的，最后一个的质量是前两者的10^8倍左右。第一个的喷流是完全非相对论性喷流，第二个是轻度相对论性喷流，第三个则是显著的相对论性喷流，其中的粒子动能超过静止能量好几个数量级（见第六章的"静止质量能量"）。尽管它们在尺度上相距甚远，但每个系统都包含一对狭窄的喷流，这些喷流携带物质以超音速飞离吸积天体。它们是自然界在差异极大的尺度和速度下创造的不同系统，却能在调整比例后做到大体相似，这说明它们的物理原理在某种程度上很简单，也就是严格限制在电磁学和引力之中，而这两门理论都没有天然的尺度限制。相比之下，地球上的一切现象都与量子力学有关，涉及的尺度要考虑普朗克常数 $h = 6.6 \times 10^{-34}$ J·s。恒星、行星、星际介质都有需要应用量子力

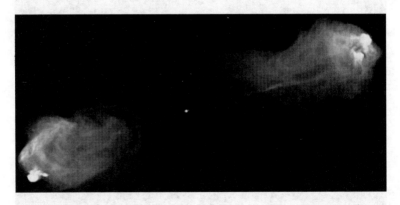

图21 射电星系天鹅座A的射电图像。细长的喷流从星系核射出，撞在超过60千秒差距外的星际周气体上

学的尺度，但喷流似乎是少有的与h无关的现象。你或许认为，既然喷流很简单，那我们一定明白它们的形成过程。但很遗憾，事实远非如此。

驱动喷流

虽然我们不完全理解喷流的形成过程，但我们自认为清楚其中的基本原理。前文提到，吸积盘应该会被不断增强的磁场线缠绕。由第三章的"日冕"一节可知，太阳表面也缠绕着不断拉伸和扭转的磁场线。磁场线在光球层外部低密度的等离子体中重连，释放磁能，引起逸出太阳的等离子体风，经过地球散入星际空间。吸积盘中心上下也一定发生着类似的事情，因此盘面上下的空间充满了炽热的气体，它们冲破系统的引力束缚，从系统中流出。但和太阳那里不同的是，这里的气体流不知为何汇聚为狭窄的喷流。也许是因为吸积盘在有规律地旋转，流出的气体被包裹在螺旋的场线中。场线缠绕着流出的气体，就像巨蟒缠住猎物要将它勒死一样。场线中的张力限制了极热气体向垂直轴线方向扩张，因此气体只能沿轴线移动，并在过程中加速、冷却。于是就形成了一道运动速度比其中的音速还快的气体窄柱。

上面描述的过程一定在不同的半径上同时发生——这是由吸积盘动力学与尺度无关的本质所决定的结果。半径大时，吸积盘更冷，转速更慢。因此可以预见，大半径下加速的气体，最终的流速会比小半径时的更慢。所以，可以认为喷流呈现嵌套结构，快速运动的中心被一层层流速逐渐递减的气柱包裹。尽管吸积盘内的气体达不到轻度相对论性的速度（见第六章），但中子星和黑洞吸积却能产生极端相对论性的喷流。我们不知道大自然是怎么实现这个奇迹的。

高效喷流

吸积盘上下的喷流以嵌套形式出现,因而本章开头"吸积盘的基本动力学"中引入的吸积盘模型需要彻底修正。那个模型建立在这个假设上:流过吸积盘内任何半径处的物质流总量都相同。但如果气体在各半径处以喷流的形式流出,那么越靠近内部,物质流的量就必然越小。另外,在之前的假设中,黏性向外带出角动量的速率,也必须与物质流带入角动量的速率相同。而存在喷流时,物质流穿过半径 r 被带走角动量的方式就不再只有黏性一种。喷流在 r 之内的每一点都会带走角动量,因此 r 处的黏性角动量流就比没有喷流时更小。因为角动量流占据了 r 处输入热量的四分之三,喷流的出现会使吸积盘变冷,辐射变弱。实际上,吸积盘的输出能量从热(形式为辐射)变成了机械能(形式为喷流的动能)。观测结果表明,这种转变几乎非常彻底,因而吸积盘释放的能量几乎全部转化为动能。

时域天文学

我们已经研究了达到稳定状态的吸积盘。然而,实际吸积盘的光度却往往会反复变化。我们可以监测整个系统的**光变曲线**,也就是光度的时间函数,从中了解吸积盘和它所在系统的许多信息。如果可以获得不同时期的光谱,还可以分析出更多的内容。

在许多系统中,大量气体通常会被抛入吸积盘的一小块区域中。这些物质随即因为转速差异而拉成一道致密的环。稍长一段时间后,黏性将环内沿的角动量带向外沿,使环的内沿向内移动,外沿向外移动。因此,原先致密的窄环变成了宽带。在状态稳定的吸积盘中,不同半径处的温度并不相同。这里也一样,环内侧的温度升高,外侧温度下降,因此整个环辐射的光谱不再类似黑体辐射谱,而逐渐演变为类似稳定吸积盘的光谱。这个过程需要的时间与黏性强度成反比,而推导稳定吸积盘的性质时并不需要知道黏性强度。

在某些系统中,光度突增是出于上述原因,由气体团进入吸积盘引发。在另一些系统中,光度突增由黏性强度突变引起。黏性低时,物质要花很长时间才能穿过吸积盘。待吸积盘状态稳定后,在给定吸积速率下,吸积盘内的气体密度较大。而黏性大时,稳定后吸积盘内的气体密度变低,但光度不变。如果稳定的吸积盘的黏性突然变大,就会使物质流出吸积盘、坠入星体的速率超过物质进入吸积盘的速率,引起光度突增,然后在高黏性状态稳定后回落到原先的程度。

那么,黏性高低变化的原因是什么?对此我们还不太了解,但其中的机制可能与黏性来源于吸积盘中的湍流,而湍流本身又由黏性驱动有关。因此,在低黏性状态下,湍流的漩涡很小,由此产生的黏性较低;在高黏性状态下,大漩涡产生的黏性也更高。

许多低质量的**X射线双星**常在发射高光度软X射线谱和低光度硬X射线谱的状态之间转换。也就是说,这种包含一颗吸积黑洞或中子星的系统,会在低-硬和高-软两种状态之间切换。系统由高-软转为低-硬时,往往会喷射相对论性的喷流。我们认为,当靠近吸积盘中心的气体沿转轴喷出,在吸积天体周围留下一块低密度区域时，就会发生上述转换。高-软状态下，致密的中心区域像黑体一样辐射；而在低-硬状态下，气体太过稀薄，本身无法生成大量光子。不过，它可以通过**逆康普顿过程**（也就是相对论性电子与光子相撞）为系统的光度出力。这就像踢任意球时，球员的鞋为球提供能量。同样，电子也能大大增加光子的能量。由此，红外线光子可以转变为X射线甚至γ射线光子，天体光谱也因此变"硬"。偶尔喷射喷流的X射线双星叫作**微类星体**。

喷流能使微类星体在射电频率的光度增加两个数量级——我们会在第六章的"静止质量能量"一节探讨其中的物理机制。总之，在射电频率上，这些光源的变化比X射线源还剧烈。当它们处于**射电强**状态时，射电辐射由喷流主导。

我们现在知道，类星体由黑洞驱动，这些黑洞的半径R_s是某些X射线双星中恒星级黑洞半径的10^8倍。这两种系统的特征速度相同，能到达光速的很大占比。因此，可以推知类星体发生变换的时标也是X射线双星的10^8倍左右。例如，X射线双星用时一秒，类星体就要三年；X射线双星变换一年，类星体就要一亿年。因此，在我们短暂的一生中，就别想观测到类星体变化了。但我们可以找出射电弱和射电强的类星体各有多少。其实，在搞清楚类星体和微类星体的关联之前，我们把类星体分为射电弱和射电强两类，其中射电弱占九成以上。这个比例与微类星体的射电弱时间占比一致。

在电磁谱图蓝色的一端（根据吸积天体的性质，可以是X射线或紫外线），许多吸积系统的亮度会以某个特征频率闪烁，这就是**准周期光变**。因为准周期光变集中在谱图的蓝光一端，这部分由吸积盘内沿的辐射主导，所以我们认为它的特征频率应当是吸积盘内沿的轨道频率。因此，由光变可知吸积天体的性质。

微类星体准周期光变的特征时标为毫秒，因此类星体的对应时标是天。这种快速涨落的振幅很小。若时标增大几百倍，微类星体的X射线光度能发生可观的变化。对于类星体，出现类似涨落的时标是年。这些涨落具有很重要的判定价值。

例如，可以通过它们估算类星体黑洞的质量。吸积盘的光度增大时，黑洞外某距离处的气体云会被吸积盘内侧的电离光子激发，增强可见光和紫外线的辐射强度。但吸积盘光度提升与发射线增强之间有个延迟时间 T，因为电离辐射需要花时间穿过类星体到轨道气体之间的距离 $r = cT$。辐射气体的轨道速度 v 可以由发射线的宽度估算。因此，由圆周运动的速度公式 $v^2 = GM/r$ 可以得到黑洞的质量为 cTv^2/G。

若与很强的引力透镜（见第六章的"引力透镜"）联合，类星体吸积盘光度的涨落还可以用来确定宇宙的尺度，以及寻找暗物质块。

第五章
行星系

天体物理学始于牛顿对太阳系动力学的研究（表1），而解释太阳系如何形成并演化到现今状态，则是今日天体物理学的研究前沿。

1995年，米歇尔·马约尔和迪迪埃·奎洛兹宣布发现了一颗行星，它环绕着与太阳相似的飞马座51做轨道运动。自此之后，人们又发现了大约一千个系外行星系。研究它们如何形成、演化为当前状态，也深刻影响了我们怎么思考自己的星系，思考自己在宇宙中的位置。我们对行星系形成和演化的认识在飞速发展，但我们依然不知道太阳系有多么不同寻常。

行星系的动力学

牛顿证明，如果各大行星单独在太阳的引力场中运动，那么它们的轨道都是椭圆（图22）。他认识到，完全理解太阳系的动力学是条漫长的旅途，上面的结果只是开了个头。因为行星的质量不是零，我们还要考虑行星间的引力影响。在接下来的250

表1　太阳系。符号 $M_⊕$ 代表地球质量，P_J 是木星周期。虽然2006年国际天文学联合会取消了冥王星的行星身份，认为它只是柯伊伯带里一颗较大的天体，但表中仍然列出了它的数据

行星	$M/M_⊕$	a/AU	e	i	周期	$P:P_J$
水星	0.055	0.387	0.206	6.34	0.241	1 : 49.2
金星	0.815	0.723	0.007	2.19	0.615	1 : 19.3
地球	1	1	0.017	1.58	1	1 : 11.9
火星	0.108	1.524	0.093	1.67	1.881	1 : 6.31
木星	317.8	5.203	0.049	0.32	11.86	
土星	95.15	9.582	0.056	0.93	29.46	2.48 : 1
天王星	14.54	19.19	0.047	1.02	84.02	7.08 : 1
海王星	17.15	30.07	0.009	0.72	164.8	13.9 : 1
冥王星	0.002	39.26	0.245	17.1	247.7	20.9 : 1

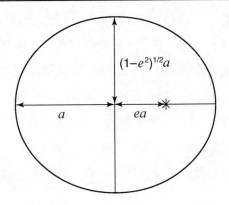

图22　偏心率 $e = 0.5$ 的椭圆轨道。主轴长度分别为 a 和 $\sqrt{1-e^2}\,a$。引力中心标记为星号，距离椭圆中心 ea

年里，这个目标一直是数学物理的前沿问题。之后，于尔班·勒威耶（1811—1877）的工作达到了一个高峰：他证明牛顿力学在计算水星轨道时有轻微的反常。1916年，在验证爱因斯坦新提

出的广义相对论时,有一条论据就是它能自然地得到那个反常结果。

最近二十年有两个条件得到了发展,这重新激起人们对行星动力学的兴趣。一是运算快且相对廉价的计算机变得容易获得,我们由此可以计算完整的运动方程,预测几十亿年的演变。二是我们发现了系外行星系,它们与太阳系往往截然不同,引得天体物理学家好奇背后的原因,以及它们能给我们带来什么启示。

受摄行星

由于行星的质量远远小于太阳,研究行星动力学最自然的方式无疑是**摄动理论**:将各行星放在各自都没有质量时所在的轨道上,然后考察某颗行星在其他行星的作用下,轨道会如何变化。在这个体系中,行星的轨道始终是牛顿式的绕日椭圆轨道,但被考察的行星轨道会在其他行星的摄动下缓慢变化。

能够确定椭圆轨道的关键量有:**半长轴** a,它决定了轨道的能量($E = -GM/2a$);**偏心率** e,它调节椭圆的形状(图22);**倾角** i,它是椭圆平面与**不变平面**的夹角,其中不变平面是个假想平面,由太阳系的角动量确定(图23)。表1列出了太阳系各行星的上述数值。我们用摄动理论计算它们如何随时间变化。

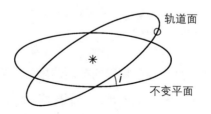

图23 倾角 i 是不变平面与行星轨道平面的夹角

角动量是吸积盘动力学（见第四章的"吸积盘的基本动力学"）中一个重要的量，它在行星动力学中同样如此。对于确定的半长轴a，角动量

$$L = \sqrt{1-e^2}\sqrt{GMa} \tag{5.1}$$

在偏心率$e = 0$时最大，此时轨道为正圆；在e趋于1时减小到0。

我们假设各行星的质量分散在轨道上，于是每条轨道化为密度稍有起伏的椭圆线，它在距离太阳最远处密度最大，因为行星在此处的运动最慢（图24）。这些线在引力作用下相互吸引，又因为它们都是椭圆且不在一个平面上，所以会像图24那样相互施加扭矩（第51页）。扭矩代表了角动量的变化率（见第四章的"吸积盘的基本动力学"），因此行星间会交换角动量。用椭圆线代替行星是个合理的近似处理，在此情况下，行星不会交换能量，所以各行星的半长轴a保持不变，偏心率e则不断变化。

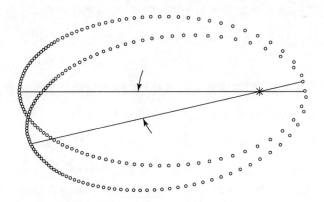

图24　两个轨道相近的行星位置图，用时间间隔相同的一百个点表示。我们用密度正比于图中点密度的椭圆代替行星。椭圆会受到另一个椭圆的引力作用，箭头标示了扭矩的方向

如果各行星的质量可以忽略不计,那么每个椭圆的长轴方向将维持不动。如果行星系的质量与不变平面内轴对称的薄圆盘等效,那么椭圆的长轴就会缓慢旋转,方向与行星在椭圆轨道上运动的方向相反。长轴的这种向后运动叫作**进动**。

在这个轴对称的行星系统中,每颗行星都有自己的进动频率。于是,某行星向另一行星施加的扭矩会来回变换方向。因为两个椭圆的长轴存在夹角,随着时间推移,有时行星1会向行星2传递角动量,有时则是行星2向行星1传递角动量,两种情况的时间相当。于是,各行星轨道的偏心率会发生轻微振荡,但也就仅此而已了。特别的是,两个椭圆长轴的夹角会一直变大。

这时会出现一种情况:两颗行星的进动频率 Ω_1 和 Ω_2 会发生**共振**,即满足

$$n_1 \Omega_1 \simeq n_2 \Omega_2, \quad (5.2)$$

其中 n_1 和 n_2 是小整数。行星1可以长时间向行星2传输角动量,使二者的偏心率都发生巨大改变。行星偏心率变化后,进动速率也随之变化,因此两椭圆长轴的夹角增速变得不均匀。实际上,它不再一味变大,而是开始振荡。我们把椭圆夹角的这种运动称为**天平动**,把不满足共振条件(5.2)的运动称为**圆环动**。

要想理解涉及微扰的各个物理分支中的许多现象,上面描述的共振可谓非常关键。因为微扰只有以同样的方式作用很长时间,才会产生重大的效果。如果没有共振,扰动就会不停变化,时间上的平均效果为零。共振使微弱的干扰有机会以同一方式作用很长时间,由此产生重大的变化。

我们将行星用椭圆代替后产生的共振称为**长期共振**,以此

区别行星沿椭圆运动的频率 Ω_ϕ 之间的更基础的共振——在椭圆线模型下,这个共振被抵消掉了。若存在小整数 n_1 和 n_2,使

$$n_1 \Omega_{\phi 1} = n_2 \Omega_{\phi 2}, \qquad (5.3)$$

就称两行星在做**平运动共振**。平运动共振下的行星既会交换角动量,又会交换能量。因此,除偏心率之外,它们的半长轴 a 也会改变。

行星的诞生

年轻的恒星外围往往环绕着吸积盘。随着恒星质量增大,它的核心温度也不断升高(见第三章的"恒星的形成")。恒星的质量如果超过 $0.08\,M_\odot$,就会发生核燃烧(见第三章的"核聚变")。之后恒星的光度增大,开始向周围的吸积盘辐射热量。受热的气体于是摆脱年轻恒星的引力,回到星际空间当中。吸积盘中的尘埃与颗粒太重,无法脱离。因此,在恒星加热吸积盘的过程中,盘内的尘埃相对气体的比重增大。尘埃在吸积盘里越积越多,它们相互碰撞融合成更大的颗粒。最终,大型的尘埃团直径达到几千米,引力也足以显著改变附近气体和尘埃的运动。**小行星**由此形成。

之后,小行星又不断碰撞、融合成越来越大的小行星。最大的几颗在自身引力作用下变成漂亮的球形,内部结构则因密度高的物质下沉、密度低的物质上浮而呈辐射状。这些球体就是行星核。

如果在比较早的时候就形成了质量很大的核,此时它所在轨道的气体还没有被全部驱散,它可以将一部分气体束缚在自

己的引力场中。这就是木星、土星、天王星、海王星这几颗大质量外行星的成因。而岩质内行星即水星、金星、地球、火星则没能捕获足够的气体，因为它们的引力场不足以在太阳系较为温热的内部留住氢和氦。

行星系的演化

早先，尘埃与颗粒还没有形成能够影响运动的引力场，气体和尘埃在行星系的不变平面中以近乎圆形的轨道运动。之后，每颗年轻行星的引力场都会在行星盘里形成螺旋波（图25）。行星会被附近密度增加的螺旋吸引。内部区域将它拉向旋转的方向，因此它从这里获得角动量；外部区域则将它往回拉，于是它将角动量传向那里。详细计算表明，行星损失的角动量比获得的更多。如果行星的质量非常大，这样的角动量交换就会使行星附近形成一道密度很低的环。这道**真空环**中的物质要么被扫入行星之中，要么被推到环的边缘之外。

做轨道运动的天体一旦失去角动量，就会向内移动。因此行星慢慢靠近恒星。真空环内沿之内的物质因为将角动量传递给行星而向内移动，但只要气体还够多，黏性向外传递的角动量就能补充气体向行星传递而损失的角动量。同样，在真空环外沿之外，黏性带走了行星传递的角动量，避免外沿向外移动。因此，行星和真空环会缓慢地向内移动并**引导**气体盘中的物质，使它们处在真空环两侧的安全距离之外，与行星保持良好的引力接触。

各行星向内漂移的速率不尽相同，因此一颗行星的真空环内沿可能会撞上另一颗行星的真空环外沿。数值模拟显示，嵌

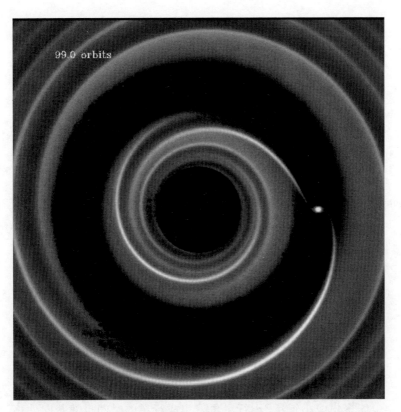

图25 绕中心（不可见）恒星运动的行星（位于右侧中心），在周围的气体盘中产生螺旋波。行星通过这个波从轨道内的气体获得角动量，同时向轨道外的气体输出角动量。角动量转移在行星周围形成一片密度趋近于零的区域

有行星的气体盘中，两颗行星很可能会进入平运动共振状态[见(5.3)式]，并能交换能量和角动量。最终，双方在交换下锁定在平运动共振状态。于是靠内的行星会从它的真空环内侧的尘埃和气体中获得能量和角动量，靠外的行星则会向它的真空环外侧的尘埃和气体输出能量和角动量。靠内的行星向靠外的行星输送充足的能量和角动量，使二者维持平运动共振。行星单独

与所在的吸积盘交换能量和角动量时,它总是会有所损失,向内漂移。但一旦两颗行星通力合作,它们就不一定会输,可能会缓慢地向外或向内移动。

年轻的太阳系

科学家认为,土星和木星早年曾有过前文描述的那种碰撞,随后二者进入2∶3的平运动共振(三木星年与两土星年等长),并且在合作下大幅阻止了向内漂移的进程。它们之外的下一颗行星,姑且称其为冰巨星1,向内漂移遇到几乎稳定的木星-土星组,也进入平运动共振,可能是与土星保持2∶3的关系。随后,三颗锁定的行星以非常平缓的速度漂移。接着,下一颗行星——冰巨星2到来。它与冰巨星1进入平运动共振,共振比大概是3∶4。此时,这四颗行星通力合作,维持在几乎不变的轨道上,而年轻的太阳还在吹散吸积盘中的气体。

由于环绕恒星的轨道运动周期正比于$r^{3/2}$(r是半径),因此,尘埃聚集成小行星、小行星再聚集成行星的时间也随着向外移动而延长。在大约20天文单位外的地方,这个过程在太阳吹尽气体后仍未完成。所以,冰巨星2之外不再有冰巨星3,只有大约1 000颗冥王星大小的天体和无数小行星。气体吹散后,小行星不再受到阻碍,它们的偏心率开始受到冥王星大小的天体的影响。这群小行星和"冥王星"们环绕在四颗锁定的行星之外,但没有延伸到冰巨星2的真空环。如今的柯伊伯小行星带就是它们的遗存,我们称它们为**原柯伊伯带**。

冰巨星中的一颗(可能是1)的轨道偏心率有些高,能够与原柯伊伯带内沿$r<20\,\mathrm{AU}$处的小行星交换能量和角动量。如

果它没有与其他行星锁定在平运动共振状态,那么它就会因损失能量而进入半径更小、形状更圆的轨道。但由于锁定,它的轨道偏心率反而变得更高。一段时间内,它的偏心率会持续增长,然后突然间,四行星系统的长期共振使冰巨星1的偏心率下降,各行星的角动量交换打破了平运动共振条件。之后,各行星不再交换能量,于是任何角动量损失都会导致偏心率升高[见方程(5.1)]。两颗冰巨星的偏心率迅速增大到很大程度,它们穿过各自的轨道,甚至可能穿过了土星轨道。此时太阳系的情况极其危险,因为偏心率极大的行星会诱使其他行星也进入偏心轨道。如果木星处在高偏心率轨道,那么很快它就会将其他行星驱赶进太阳或彻底逐出太阳系。

之后我们会看到,这样的灾难可能在许多行星系发生过。我们可以说是运气好,还有原柯伊伯带起了预防作用。随着两颗冰巨星的偏心率越来越大,它们穿过原柯伊伯带与小行星和"冥王星"们近距离接触。驱散小天体使两颗冰巨星的偏心率降低,太阳系逐步形成今天的布局。如今,海王星与天王星保持着1∶2的平运动共振,轨道偏心率低,半长轴 a = 30.1 AU,深入原柯伊伯带。天王星的轨道也可能在原柯伊伯带之内(表1)。根据某些关于四行星系统的模拟结果,共振状态破坏后,冰巨星1最终的轨道比冰巨星2更小,在另一些结果中则更大。所以,我们不知道哪一颗是海王星。

两颗冰巨星穿过原柯伊伯带时,摧毁了其中的大量天体,因此如今的柯伊伯带只包含大约 0.07 $M_⊕$ 物质,与我们认为它最初拥有的大约 40 $M_⊕$ 物质相差甚远。上千颗"冥王星"中只留下一颗,大量小行星被赶出太阳系。不过,其中有不少会在某个时候穿过原柯伊伯带，被水星到土星的各个行星驱赶，在它们的表面撞出陷坑，降低了它们的偏心率。其实，小行星撞击月球的频率可以根据陨击坑的样式确定。我们在提出上述太阳系的演化过程之前很久，就知道太阳形成的7亿年后（太阳形成于46亿年前），月球经历了一次**后期重轰炸**。这段高密度小行星时期的另一个遗存可能还有木星的**特洛伊小行星**，它们与木星共处同一条轨道，但在太阳的另一边。我们认为木星就是在上述时期捕获了它们。

秩序、混沌和灾难

牛顿给我们留下了一件预测未来的利器：微分方程（第2页）。以描述当前动力系统分布的数值为初始条件求解方程，可以由结果读出系统未来任意时刻的预测结果。遗憾的是，在20世纪，这个方案常常不起作用。问题在于，求出的结果对初始条件**极其**敏感。亨利·庞加莱（他的一个堂兄弟领导法国参加了1914—1918年的第一次世界大战）在20世纪初率先发现了这个问题，但直到1960年代电子计算机能够运算一般动力学系统的微分方程后，这个问题的深度才显露出来。在这一问题中，太阳系再次成为最简洁、最让人印象深刻的案例。

我们很容易写出掌控太阳系行星运动的微分方程，而天文观测能让初始条件达到很大的精度。但在这样的精度下，我们也只能预测未来至多4 000万年的行星分布——如果初始条件有变但在观测误差内，预测5 000万或6 000万年后的结果也会大不相同。如果想让预测6 000万年的结果与4 000万年的同样准确，初始条件的精确程度就要提高100倍。例如，若要当前各行星位置的观测误差不超过15米,并以同样的精度预测6 015万年后的结果,那么当前位置的误差就必须在15毫米以内。这里,我们碰到了一个严峻的限制。照此看来,我们对太阳系当前分布的观测结果,还不能预测大约6 000万年后各行星的位置。这在太阳系46亿年的年龄面前不过是弹指一挥间,实在令人泄气。

在这种情况下,我们能做的只有用微分方程计算未来最有可能的分布。做法是随机选择符合当前分布的初始数据,计算对应的结果,由此预测各行星会出现在什么位置。当前分布的采样中,如果有许多得到非常接近的结果,那它便是最有可能的未来分布。

太阳系微分方程的解有一个重要特征:某个变量(比如水星轨道的偏心率)在小范围内涨落几百万年后,会突然变换到完全不同的范围内。这种现象表明共振对系统的动力学有重要作用:在某个瞬间,系统满足共振条件,其中的能量流动也发生变化,因为很小的扰动会在千百次的循环中积累,产生巨大的影响。如果从一个与上一分布有微小差异的分布开始积分,共振条件就不再满足,变得要么提前,要么延迟,得到的解也大不一样。

共振的重要作用使得解对于物理定律的细节极度敏感。我们将在第六章介绍爱因斯坦的相对论,探讨它在天文学上的一些应用。这里提前给出相对论的重点,也就是v^2/c^2这个量,其中v是一般速度,c是光速。对地球而言,二者比值大约是10^{-8},可以说非常非常小,即使换成水星也不超过它的2.5倍。然而,雅克·拉斯卡和迈克尔·加斯蒂内奥主导的实验却表明,如果不对太阳系的牛顿理论做出这么微小的修正,我们恐怕就不复存在了。

拉斯卡和加斯蒂内奥让太阳系从当前分布开始演化,得到了两套解。两套计算使用的初始条件都符合最佳观测数据,由此算出太阳系的分布解。其中一套用不含爱因斯坦微小修正的牛顿理论,计算了未来50亿年的解,另一套则考虑了爱因斯坦的修正。他们发现,计入爱因斯坦修正后,水星进入偏心率 $e > 0.7$ 的情况只占所有解的1%;没有计入爱因斯坦修正的2 500个解中,只有1个的水星轨道 $e < 0.7$。如果水星的 $e > 0.7$,地球将面临严重的后果,因为偏心率高的水星会很快拉高金星的偏心率,金星又会让地球高度偏心,地球又让火星偏心率升高。于是,剧变接踵而至:未修正的牛顿方程的2 500个解中,水星与金星相撞的有86个,与地球相撞的有34个。即便几颗内行星没有彼此碰撞,它们也会被拉进太阳内或甩出太阳系。

因此,我们似乎身处险境。如果没有爱因斯坦对牛顿方程做出的那微小的修正,地球几乎不可能再为我们提供庇护。即便有爱因斯坦的修正,也无法保证未来大约8 000万年后的安全。无论如何,多亏有了广义相对论,地球才极有可能幸存到40亿年之后,直到被膨胀的太阳吞没。

爱因斯坦的修正让木星和水星的共振失谐,然后生命才得以存在。因为它们之间的是弱共振,只有各项条件都恰好满足才能产生影响。而爱因斯坦的修正使它难以达到要求。

系外行星系

第一个明确探测到的系外行星系由米歇尔·马约尔和迪迪埃·奎洛兹在1995年发现。如今,我们已经知道了一千多个行星系。根据这些已知系统的结构统计数据,可以一窥行星系演化的奥秘：这就好比宇宙给演化方程计算了大量的解，以此保证结论绝对正确。

我们看到，太阳系顶住两次大威胁，留下了八颗轨道近乎圆形的行星。第一次威胁在它诞生7亿年后，大行星的紧密结构被破坏。即使到今天，如果将内太阳系的物理性质改变一亿分之一，它也会陷入危险。

因此，也无怪第一批被发现的行星系都像是经历过上述灾难一样，含有一颗类似木星的行星，沿着偏心轨道以相对较短的周期运动。然而，如果考虑如下事实，这个发现也并没有那么重要：发现第一批行星系的观测技术更容易找到带有短周期大行星的星系。它的方法是监测恒星速度的周期变化，这个速度是相对恒星与行星的质心而言。行星质量越大、越靠近恒星，速度变化程度也越大，因此它们的信号也越容易从噪声中检测出来。

之后，科学家使用了一种完全不同的方法来探测行星，由此找到了包含多颗行星的行星系，这些行星都以近乎圆形的轨道运动。这项技术靠的是监测恒星的亮度。在行星经过恒星与地球之间的区域时，可以发现恒星的亮度有小幅下降。这种方法发现的行星系都处在非常边缘的位置，达到精度要求的数据只能从太空中获得。2009年5月，美国航空航天局发射了开普勒卫星，用来监测天空中一小片区域的恒星。之后的四年里，它准确地找到了将近一千个行星系，并列出了一张包含几千颗恒星的列表，这些恒星都有存在行星的迹象。评估这些数据是如今非常活跃的研究领域。

第六章

相对论天体物理学

我们熟悉的牛顿物理学世界是相对论物理世界的近似物，在速度明显小于光速时很方便，效果也不错。但天文学家发现，许多类型的天体突破了低速的限制，因此我们需要用完整的相对论为它们构建模型。

我们会在"狭义相对论"一节中概括相对论的主要结论，现在我们需要知道其中最著名的那条：$E = mc^2$。这个方程说的是，根据相对论，粒子的能量不只是牛顿物理中那样，是它的动能加上势能，还要另外加上它的**静止质量能量** $E_0 = m_0 c^2$，其中 m_0 是粒子静止时的质量。静止质量是粒子的特征量，不会变化。反过来，质量 $m = E/c^2$ 会因为对粒子做功或粒子对外做功而改变：电子在斯坦福直线对撞机之类的加速器中加速时，质量可以增大200倍左右。比值 $\gamma = m/m_0$ 被称为**洛伦兹因子**，定量表示粒子的相对论性程度。以100千米每小时运动的汽车，洛伦兹因子与1的差值非常小（$\gamma - 1 = 5.8 \times 10^{-13}$），所以明显不具有相对论性。一般来说，任何满足 $\gamma - 1 \gtrsim 0.1$ 的物体都可以被视为在做相对论性运动。

下面列出一些需要使用相对论的情况。

射电星系 我们从射电星系探测到的大部分辐射由洛伦兹因子约为 10^5 的电子产生。这些高能电子大多随机地运动着,但射电星系的中心通常会有喷流,等离子体在其中整齐地流动,它们的动能是静止质量能量的好几倍。

微类星体 这些天体相当于袖珍版的射电星系:驱动微类星体的黑洞质量是太阳的好几倍(第68页),而射电星系中心的黑洞具有几亿太阳质量。质量上的缩减使得物理现象的规模也变小,涨落更频繁,但不会改变特征速度。因此,对射电星系很重要的相对论,对微类星体也同样重要。

γ 射线 电子的反粒子——正电子,本质上也是和相对论有关的现象,最初由 P. A. M. 狄拉克在尝试整合量子力学和相对论的过程中预言。电子与正电子湮灭时,二者的所有能量都会转化成光子——通常是两个光子。如果两个粒子是非相对论性的(因此 $\gamma \simeq 1$),每个光子就具有电子静止质量能量,即 511 千电子伏。γ 射线望远镜在银河系中心方向探测到了它的谱线,暗示那里存在高密度的正电子。

1963 年,英国、美国和苏联签订协议,禁止在大气层进行核试验。冷战期间,美国和苏联双方互不信任,都秘密发射了卫星,侦测核试验发出的 γ 射线。结果出乎所有人的意料——他们检测到了**大量** γ 射线暴。这些射线暴会持续几秒到一分钟不等,如果是核武器产生的,那也太过频繁了。

双方的军事专家秘密研究了这些数据很久,之后才得知对方也观测到了这些现象,同时也明白它们来源于宇宙。1973 年数据公开,这回轮到天文学家苦思了。这些事件似乎在天空中均匀分布,表明它们的来源与银河系的恒星无关(大部分X射线源与恒星有关)。这些射线源要么在太阳附近大约0.1千秒差距之内,要么分散在一个比银河系还大的圆柱之内。但它们的时标很短,难以归结为活跃的星系,而且也没人认为太阳附近有能够服众的信号源。1986年,玻丹·帕琴斯基大胆提出假设,认为它们虽然时标很短,但与我们的距离**确实**非常远,可能与某种恒星爆发有关。1997年,他的猜想得到了证实。当时,威廉·赫歇尔望远镜拍下了一个刚发现的射线暴的邻近区域。我们发现,这个事件的快速衰退光学**余晖**出现在一个遥远的星系中。自此之后,我们就常常检测到光学余晖,也有了背后天体的光谱。这些数据确定,许多γ射线暴的确与恒星爆发有关。我们还从中发现γ射线暴的来源有好几种。对此,我们的理解尚不完备。可以确定的是,相对论在解释这些极端天体时发挥了重要作用。

宇宙射线 地球一直处在相对论性粒子的轰炸之中——探测这些粒子是高能天体物理学最早的分支。所幸,这些危险的粒子大多会先与大气层高处的氧原子和氮原子相撞,氧、氮原子核因冲击而碎裂,向下飞散,又撞进其他原子核里。因此,一个高能粒子进入大气层就能打出一场宇宙射线之雨。

轰击地球的粒子具有各不相同的能量,不过其中的低能粒子比高能粒子要多得多。能量最高的粒子,即使在接收面积最大的探测器中也难得一见。总之,目前检测到的粒子中,能量最高的在 10^{20} 电子伏左右。如果它是电子,γ约为 10^{14};如果是质子或中子,γ是 10^{11} 左右。功率最高的粒子加速器都远远达不到这个能量。日内瓦的大型强子对撞机目前能将质子加速到 $\gamma \simeq 10^3$。

中子星 中子星表面的逃逸速度只有$c/2$左右,所以它们的相对论性不高。但我们可以对它们进行精确测量,因此轻度的相对论效应也可以精准定量,这有力地验证了相对论。其中特别重要的是**赫尔斯-泰勒**双脉冲星PSR B1913 + 16。这是一对彼此旋转的中子星,轨道偏心率$e = 0.62$,周期7.75小时。这对双星由约瑟夫·泰勒和他的研究生拉塞尔·赫尔斯在1974年发现,之后便一直受到密切观测。此后,天文学家又发现了几对类似的双中子星系统,但因为观测时间还不长,所以对相对论的影响较小。

X射线源 第四章的"吸积盘"一节提到,黑洞吸积盘最内侧的半径处非常炽热,因此辐射以X射线为主。这片区域太小,在X射线望远镜中的分辨率很低,但相对论效应可以修正我们探测到的X射线发射线的形状。

太阳系 地球以大约30千米每秒的速度绕太阳公转。因为这个速度大约是$10^{-4}c$,而且相对论效应比起牛顿力学只大约相差一个因子$(v/c)^2$,所以可以想见,相对论对太阳系只有非常小的影响。不过,由于我们可以精确地测量太阳系,所以数据仍需要用相对论解释,而且这也为相对论提供了严格的验证。另外,我们在第五章的"系外行星系"一节中看到,太阳系的动力学结构非常敏感,如果忽略相对论效应,那么太阳系的分布就会与我们实际见到的大相径庭。

宇宙 由于牛顿物理学存在概念上的问题,我们无法在它的框架内构建一个有说服力的宇宙动力学模型。因此,宇宙论在有了爱因斯坦广义相对论的加持后才成为物理学的分支。1960年代发现的**微波背景辐射**使我们能够研究大爆炸后仅仅10

万年的宇宙；之后又发现了类星体，它们大多位于以相对论性速度远离我们的退行星系中——这些为宇宙论奠定了坚实的实证基础。

狭义相对论

我在第4页曾说，物理学家认定自然定律应该在任何地点、任何时间都不变——如果不同地点或不同时间测得的现象有什么变化，那**一定**是运用普遍定律的条件有了某些改变。狭义相对论明确提出了一种新的不变性：物理定律应当在所有星系中都相同，与它们相互运动的速度有多快无关；物理定律在所有的飞船中也应当都相同，与飞船的速度是多少无关。

1899年，亨德里克·洛伦兹在麦克斯韦的电动力学方程组中发现了一种对称。现在，我们称它为**洛伦兹协变性**，并将它视为物理学的基本原理，但洛伦兹本人却不清楚这个新对称有多么重要的物理意义。1905年，爱因斯坦认为洛伦兹对称反映了一件事实：电磁学在任何飞船中的效果都完全相同，与飞船的速度无关。这个论断在当时引起了不小的震动，因为麦克斯韦的电磁波必须通过名为**以太**的介质传播，但以太不可能相对任何飞船都静止不动。实际上，地球"飞船"的速度每六个月就会改变大约60千米每秒，用光做实验应当能测出这样的变化。而令其他人疑惑的是，测定地球相对以太运动的实验，没有一例取得了成功。爱因斯坦解释说，根据洛伦兹对称，原则上就不可能测出相对以太的运动。现在，我们将光的介质称为**真空**，它拥有奇特的性质，不论观察者相互之间怎么运动，它在每个观察者看来都一样。特别是，没有观察者能说自己处于绝对静止，所以一切运动都是相对于另一个观察者而言的——这就是"相对论"这个名字的由来。

理解相对论的关键,是将每个物理情形拆分为一连串的**事件**。事件总是发生于某个地点和**某个时间**,因此可以用四个数确定,即位置坐标 x、y、z 和发生的时间 t。观察者 O′ 相对观察者 O 做相对运动,相比于 O 的四个数 (x, y, z, t),O′ 使用另一套数 (x', y', z', t') 来标记同一个事件。洛伦兹发现的简单规则——**洛伦兹变换**,可以在给定 O′ 相对 O 运动的速度 v 的情况下,由不带撇号的数算出带撇号的数。

洛伦兹变换有个奇特的结果,就是对 O 而言同时发生的两个事件 (x, y, z, t) 和 (x_1, y_1, z_1, t_1)(所以 $t_1 = t$),对 O′ 而言并不会同时发生,也就是 $t' \neq t_1'$。因此 O′ 会认为一个事件比另一个更早,而 O 认为的事实却是二者同时发生。同时性与观察者有关的思想强烈地违反了直觉,让人很难适应。

它是一个令人异常费解的原理——动钟变慢的基础。例如,鲍勃站在月台仔细观察呼啸而过的列车,车内坐着爱丽丝,她的腿上放着一块时钟。鲍勃认为时钟变慢了,因为那是一个运动的钟。另一边,爱丽丝也在仔细观察鲍勃的手表,并认为手表变慢了,因为它在相对自己运动。实际上,爱丽丝若检查时钟,会发现它走得分毫不差;同样,手表在鲍勃看来也精准无误。但在运动的观察者眼里,二者都变慢了。时间变慢的因子就是与列车车速有关的洛伦兹因子 γ。

μ 子的寿命

在地球上测量宇宙射线直接证明了动钟原理。μ 子是类似电子的基本粒子，它们非常不稳定，会很快衰变为其他粒子——它们的**半衰期**（样本中一半发生衰变的时间）是2.2微秒。宇宙射线粒子击中地表上方20千米左右大气中的原子核后，就会产生μ子。即使它们以光速运动，在半衰期内也只能移动大约660米。因此你或许会认为，在地表上方20千米处产生的μ子即使能到达地面，肯定也很少，跟用气球探测器检测到的数量没法比。但事实并非如此。这个矛盾可以用相对论解释。根据相对论，虽然我们的时钟走过了2.2微秒，但μ子的时钟只经过了$2.2/\gamma$微秒，所以μ子要等到我们的时钟走了2.2γ微秒后才会衰变。于是，以接近光速c运动的μ子在衰变前会移动660γ米。因此，在大气层产生的$\gamma \gg 1$的μ子，有很大的概率能到达地面。

静止质量能量

爱丽丝和鲍勃对给定粒子的能量或动量有不同的看法。因为如果粒子相对爱丽丝的参考系静止，她就会说它没有动能或动量，而鲍勃会说它有动能和动量。因此，若要由爱丽丝确定的数值计算鲍勃看到的能量和动量，就需要某条法则。神奇的是，这条法则就是洛伦兹变换。所以，如果在爱丽丝看来，给定质量为m_0的粒子的动量和能量分别是(p_x, p_y, p_z, E)，那么就可以将鲍勃相对于爱丽丝的速度代入洛伦兹变换，应用于$(p_x, p_y, p_z, E/c^2)$[而不是事件的坐标(x, y, z, t)]，算出鲍勃看到的动量和能量(p_x', p_y', p_z', E')。由这条简单的法则可知，静止质量为m_0的粒子，能量$E = \gamma m_0 c^2$。特别当$v \ll c$，也即$\gamma \simeq 1$时，有$E = m_0 c^2$，这就是爱因斯坦著名的质能方程。

根据量子力学，角频率为ω、波长为λ、沿单位矢量**n**方向运动的光子具有能量$\hbar\omega$和动量$\hbar\mathbf{k}$。其中$\hbar$是普朗克常数除以2π，$\mathbf{k}$是**波矢**，$\mathbf{k}=(\omega/c)\mathbf{n}$。对爱丽丝的$(k_x, k_y, k_z, \omega)$运用洛伦兹变换，可得鲍勃眼中的波矢$\mathbf{k}'$和角频率$\omega'$。因为光子其实是一列波，可以想见，鲍勃测得的角频率ω'相对爱丽丝的会产生**多普勒频移**，而洛伦兹变换就是计算这个频移的正确方法。它还能让我们看到光子传播的方向如何受到鲍勃相对爱丽丝运动的影响，因为它的方向由单位矢量$\mathbf{n}=(c/\omega)\mathbf{k}$和$\mathbf{n}'=(c/\omega')\mathbf{k}'$给出。

因此，不同观察者对给定光子移动的方向也没有统一意见。下面就用一个例子从物理上直观地说明他们为什么意见不一。老式的西部电影中，常有警卫在运输黄金的火车上朝劫匪开火的桥段。如果警卫射击的方向与车厢垂直，那么子弹并不会沿垂直铁轨的方向射出，因为它除了沿枪管运动的速度之外，还有随火车向前的运动。所以，如果警卫想要打中与自己的连线垂直于铁轨的目标，就要向垂直车厢方向的后方瞄准。这样，子弹沿枪管的速度就有一个分量抵消火车向前的运动。

现在，设想一群慌乱的警卫在火车上均匀地朝各个方向射击。那么，他们打出的一半子弹会沿着垂直车厢向前的方向射出，另一半沿垂直向后的方向射出。于是，如果从地面观察，会看到一半以上的子弹向前运动：火车前进使得弹幕整体也**向前成束**。

喷流

我们在第四章的"喷流"一节看到，许多天体会喷射等离子体喷流。喷流的平均速度很快，相应的洛伦兹因子γ可以达到1的好几倍。在随喷流物质一同运动的观察者看来，喷流内部经过各种过程射出的光子，在空间各个角度的分布大体均匀。从我们的角度看，这些光子就像警卫的弹幕一样，特别偏向前方。实际上，出射光子中有一半会集中在喷流中轴附近的方向，只占散布球面的 $1/(2\gamma^2)$。即使 γ 的值极小，这个比例也是很小的。图26直观展示了球面光子密度与角度 θ 的函数关系，其中 θ 是光子方向与喷流中轴的夹角。喷流速度可以忽略时（$\gamma \simeq 1$），任意 θ 的密度都是1。可以看到，即使 $\gamma < 2$，向前的光子束就已经很强了。实际上，喷流向前集中**能量**的程度比图26展示的还高，因为向前的光子受到多普勒效应影响，频率会大幅增加，因此能量也随之增加。

通常来说，容易形成喷流的天体会射出两道方向相反的喷流。其中一道一般具有朝向我们的速度分量，另一道具有远离我们的速度分量。由于喷流向前集中，朝向我们的喷流会比远

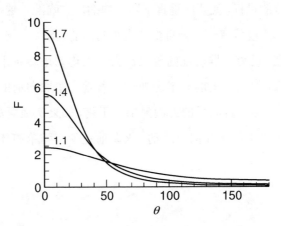

图26　洛伦兹因子 $\gamma = 1.1$、1.4和1.7时喷流出射光子的角度分布。图像表示的是各个方向上每单位立体角的光子数。对各向同性分布来说，光子数是1

离我们的更亮。又因为天体必须超过某个临界亮度才能被探测到，所以我们有时只能看到朝向我们的喷流。如果我们正好沿着靠近喷流发射轴线的方向观测，就可能出现这种情况。许多射电星系看上去就只有一道喷流。

有时，喷流会呈现相对论不允许出现的**超光速运动**。这种现象指的是喷流中的一些团块似乎以超光速的速度穿过天空。计算假设团块在天空平面内运动，那么任意垂直天空平面的速度分量只会增加它的移动距离，得到的速度也因此增大。

图27展示了相应的几何关系。两个空心圆代表同一个团块，在t'时刻，（上方的空心圆）团块离出射源较近，之后的t时刻离出射源较远。t'和t时刻团块都向地球发射了光子。下方的星形表示第一个光子在t时刻的位置。在冲向地球的"比赛"中，第一个光子抢在第二个光子之前出发，领先优势折算成距离是

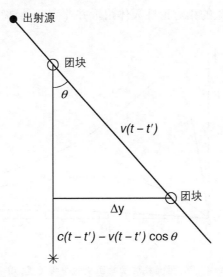

图27　超光速膨胀的几何示意图

$c(t-t') - v(t-t')\cos\theta$，因此它到达地球的时间会提前

$$\triangle t = (t-t')\left(1 - \frac{v}{c}\cos\theta\right)。$$

在发射两个光子的时间内，团块在天空平面移动了$\triangle y = v(t-t')\sin\theta$的距离。因此它的视速度为

$$v_{视} = \frac{\triangle y}{\triangle t} = \frac{v\sin\theta}{1-(v/c)\cos\theta}。$$

图28展示了0.2、0.71和0.96三个v/c值下$v_{视}$随θ变化的图像。可以看到，$v > 0.71c$时就可能出现超光速膨胀了；在$v \gtrsim 0.95c$时，$v_{视}$可以是c的好几倍。我们已经在许多射电源中观测到了超光速运动。

激波与粒子加速

发射喷流的相对论性天体附近并非完全是真空。喷流从喷

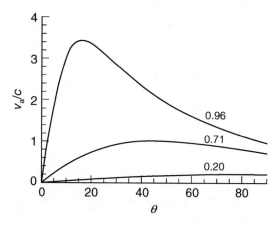

图28 以速度$v = 0.96c$沿直线运动的团块的平均速度，运动方向与视线夹角为θ

射源射出时的密度非常高，因此环境介质的密度可以忽略不计。但喷流在远离喷射源的过程中会不断扩散，使得密度下降。因此，最终环境介质会对喷流产生巨大的影响。

要想象喷流穿过环境介质的景象，最简单的就是设想你以介于喷流和环境介质的速度移动。假设喷流从你左侧飞来，介质从你右侧靠近。你所在的位置被喷流与介质碰撞产生的物质填满。这些物质非常热，因为左侧来的喷流和右侧来的介质原本都在有序运动，但到了这里却变成一个个粒子随机地横冲直撞：每个粒子都在不停地折返，但整体上看，它们并没有移动，因此它们组成的**受震激等离子体**与你保持相对静止。

受震激等离子体区域不断远离发射体并稳步增长，因为一直有新的喷流物质和环境物质从左右两侧撞来。喷流和环境粒子成体系的运动在这里变得杂乱而随机，我们把这块狭窄的区域称为**激波**。

物理学家说的激波，一般指**无碰撞**激波。可是，激波正是快速运动和慢速运动的流体相撞的地方，"无碰撞"的叫法似乎有些怪诞。它的真正含义是，改变粒子速度的是电场，这个电场作用的距离远大于粒子间的距离。因此，入射的电子、质子并不是因为与其他原子和离子碰撞而减速，它们早已在电场中经历了相对平滑的减速过程。这个电场由无数个电子和离子集体产生（图29）。

电子和离子因为巨大的质量差异（相差1 800多倍）而分离，这是电场大体上的来源。入射的电子比更重的离子减速更慢，因此正负电荷各自区域的密度不断增大。二者产生的电场一边拉住电子，一边拉住离子，慢慢使双方具有相同的平均速

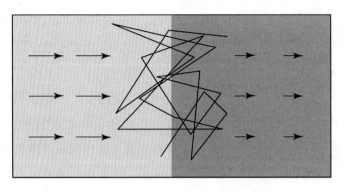

图29　由左侧冲来的等离子体撞上右侧缓慢运动的等离子体。经过碰撞，等离子体减速，密度变大，标示为较短的箭头和更深的阴影。折线代表一个高速粒子的轨迹，它经历多次散射，其间常常穿过两个区域的交界面

度。电子与离子的平均速度有差异，就会表现为电流，电流又产生磁场。再者，这个区域的电子流和离子流极不稳定，因此电场和磁场都在随时间变化。因为场随时间变化，它们会改变单个电子和离子的能量，平均来看就是将能量从离子转向电子。最开始，有序物质冲入激波，每个粒子的速度都相同，因此物质的动能集中在离子上。受震激后的等离子体**弛豫**到热平衡状态，其中每个粒子的能量正好是此时随机动能的一半。弛豫过程的关键是动能整体上由离子转移给电子。

　　受震激等离子体好比一间赌场，赌徒带来的钱在那里重新分配。上面讨论的就是这种分配对每个赌徒的影响。不过，其中一些赌徒会在一开始就小富一把，而后变得极其富有。

　　在等离子体中，粒子最重要的财富是动能。粒子运动越快，就越难偏转。快到一定程度后，粒子会直接冲出激波，向左或向右进入有序的物质流中。因为有序区域非常大，进入后的粒子**最终仍会**受到偏转，辗转回到激波之中。但回来后的粒子比离

开时更快，因为它从有序物质流中折返后的净效果是**相对**物质流改变了速度的方向。在非相对论情况下，它此时的速度是原先的速度加上有序物质流的速度。由于粒子运动得比之前更快，它很可能会冲过激波进入另一边的有序物质流。同理，之后它会再次反向，以更大的速度返回。经过这种**费米加速**过程的粒子可以获得极高的洛伦兹因子。其实，我们探测到的宇宙射线就是以这种方式加速的。

受震激等离子体非常热，是一种高压强流体，会向任意方向膨胀。左侧喷流和右侧环境介质的**冲压**使它无法向这两边扩张，但它可以向垂直方向膨胀。以这种方式沿喷流垂直方向扩散的物质最终会膨胀为包裹喷流的等离子体**茧**，如图30所示。

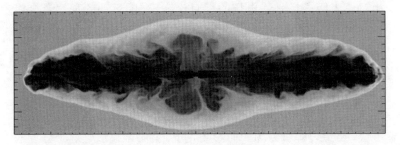

图30 计算机模拟的由喷流形成的茧。深色阴影表示低密度区域。灰色边缘表示未受干扰的星系周气体。边缘包裹的浅色带是激波压缩的气体。内侧的深色阴影表示湍流形式的极热等离子体，它们被喷流末端的激波加热

同步辐射

茧中有许多电子因为费米过程而加速，洛伦兹因子变得很大。茧内始终包含磁场，电子沿磁场的力线螺旋前进，同时发出电磁辐射。如果电子是非相对论性的，它们的辐射频率就都是**拉莫尔频率** v_L，与磁场强度成正比。如果电子是相对论性的，那

么辐射会覆盖一段频率范围,最高到 $\gamma^2 \nu_L$。这种辐射叫作**同步辐射**。出射电子的典型洛伦兹因子可以由发射谱推断。因为星际和星系际空间的磁场通常很弱,只有洛伦兹因子 $\gamma \gtrsim 10^4$ 的电子发出的同步辐射才容易被射电望远镜探测到。尽管如此,同步辐射依然很常见。

广义相对论

前面提到,狭义相对论源自麦克斯韦的电动力学方程组,但它只揭露了时空的基础对称结构。爱因斯坦相信**一切**基本物理定律都具有这种对称,而牛顿的引力论显然不满足这一条件。

引力与静电力有非常相似的属性:都与距离的平方成反比。相对论表明磁场是静电场的相对论修正,即运动的观察者会将一部分电场看作磁场。因此,在运动的观察者眼中,引力场也应当有所不同。特别是作用在物体上的完整引力,应当与物体的速度有关,就像作用在电荷上的电磁力有一个正比于速度的磁场分量。由于光子的运动速度比任何静止质量不为零的粒子都更快,要想理解引力场如何影响光子,就必须搞清楚这个分量的情况。

引力有一个静电力不具备的性质,爱因斯坦认为它触及了问题的根本:引力与质量成正比。传说伽利略在1600年前后将两个质量不同的球从比萨斜塔塔顶抛下,证明了上述性质——尽管两球质量不同,但它们却几乎同时落到地面。伽利略的实验算不上精确,一种更好的验证引力与质量相关的方法是,用不同材质和质量的摆锤制作摆长相同的单摆,然后测量单摆的周期。1891年厄缶男爵设计了一个实验,根据物体受太阳的引力以及它围绕太阳运动所需要的力的平衡状态,非常精确地证明了引力与质量成比例。爱因斯坦认为,这么精确的比例关系不可能是巧合,正确的引力论必然能得出这个结果。

爱因斯坦苦心孤诣十年,为自己的想法赋予数学的血肉。他的成果毫无疑问是人类创造的最伟大的成就之一。本质上看,他对引力做了麦克斯韦对电磁学做过的事,也就是将单独的物理现象囊括在一套连贯的数学结构之中。其中不仅包含启发了它的物理事实,还预言了全新的现象。

电磁场由电荷和电流产生。引力则由能量-动量及其流量产生。在牛顿物理学中,引力源于质量,而根据 $E = mc^2$,质量等价于能量。广义相对论则教导我们,这个观点太过狭隘,我们之所以这么想,是因为我们没有见过高速运动的大质量物体或非常高的压力,所以没有意识到动量和能量-动量流都会产生引力。我们还认为引力天然是吸引力,但广义相对论表明,它也可以是排斥力。

麦克斯韦方程组是微分方程,可以解出给定电荷和电流密度产生的磁场;爱因斯坦方程也是微分方程,可以解出给定能量-动量密度及其流量产生的引力场。不过,麦克斯韦方程组都是线性方程,爱因斯坦方程则是**非线性方程**:对于**线性方程**,只须找到简单解,然后把它们加起来就可以构造更复杂的解;对于非线性方程,两个解的和通常不是解,因此无法将简单解相加得到复杂解。

由于这个问题,我们迄今为止得到的爱因斯坦方程的解都是特定对称情况下的解。其中最早,也最为著名的是卡尔·史瓦西在1916年得到的史瓦西解。这个解描述了球状物体周围的引力场，我们在第五章的"系外行星系"一节中讨论了它在太阳系的应用。1963年罗伊·克尔将这个解推广到旋转物体周围的引力场，它对于我们解释吸积盘内部区域（见第四章的"旅途的终点"）非常重要。其他精确解描述了各式各样的均匀宇宙。在这个约束颇多的解集中，有一个精确解可以用摄动理论（参见第五章的"行星系的动力学"）拓展为近似解，广义相对论在天文学方面有许多应用会用到这个方法。自2000年以来，我们在爱因斯坦方程的数值解领域取得了多项重大发展。它们一方面得益于可用的计算机算力稳步增长，但主要还是因为我们更懂得如何去运用这些方程。

弱场的引力　牛顿引力论的精度需要满足两个条件：其一，引力场必须较弱；其二，速度远小于c。第二个条件在实践中更加苛刻，尤其是光子不可能满足。天文学的场通常很弱且与时间无关。因此牛顿的引力势可以完整描述它的引力场。例如，在距离质点M外r处的引力势就是$\Phi(r)=-GM/r$。

引力红移

假设我们要测量一个致密天体（比如白矮星）表面的原子发出的谱线频率。我们先站在$\mathbf{x}_o$处，且可以清楚听到致密天体附近$\mathbf{x}_c$处，相对我们静止的原子钟的滴答声。假设原子钟每秒响一次，那么我们接收到的两次光脉冲的间隔就是

$$t_o = \sqrt{\frac{1+\frac{1}{2}\Phi(\mathbf{x}_o)/c^2}{1+\frac{1}{2}\Phi(\mathbf{x}_c)/c^2}}\ \text{s}.$$

因为原子钟比我们更靠近致密天体,所以 $\Phi(\mathbf{x}_c) < \Phi(\mathbf{x}_0)$,于是 $t_0 > 1$ s。因此,我们会认为原子钟变慢了。将原子钟换成振动的原子,可以测得它的振动频率是 $v_0 = 1/t_0$,比固有频率 $v_c = 1/t_c$ 更低(其中 t_c 是在 $\mathbf{x}_c$ 处测得的两次滴答声的间隔)。这种现象就是**引力红移**。

引力透镜

光穿过玻璃和水的速度比它穿过空气的速度更慢,因此光在进入玻璃或水中时会弯曲。这个现象通常用**折射率** n 量化,它的定义是:在折射率为 n 的介质中,光速是 c/n。

如果使用普通的笛卡尔坐标 (x, y, z),且假设 x 轴上两点 x_1 与 x_2 的距离 $s = x_2 - x_1$,那么引力场会使真空具有折射率

$$n = 1 + \frac{|\Phi|}{2c^2}。 \qquad (6.1)$$

在我们采用的相当自然的距离定义下,$\Phi \neq 0$ 时,光在真空中传播的速度似乎会比光速更慢。其实,光总是**以光速**运动,上式给出较低的值,是因为我们低估了 x_1 和 x_2 的距离。但设想光子在 $\Phi(\mathbf{x}) < 0$ 时速度小于 c 也很有用处。

这是因为,它能让我们运用光学知识预测光会怎样被引力场偏转。凸透镜会减慢光子的速度,使经过透镜中心的光子比远离中轴、经过更薄镜身的光子更慢,从而使平行光束汇聚在焦点上(图31上)。实际上,凸透镜的外形设计就是使光子从远处光源到达像的时间与光子穿过透镜时到轴线的距离无关,这就是**费马最短时间原理**。

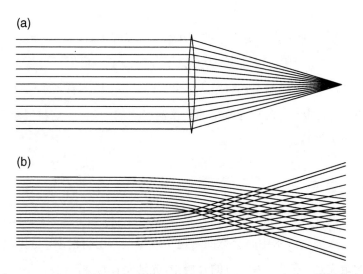

图31 凸透镜的外形设计使它能让原本平行的光线在穿过后经过焦点(a)。引力透镜(b)会使原本平行的光线汇聚,但不能使它们经过同一个点。透镜天体质量极大,远远超乎我们想象,所以我们很容易看到它们引起的折射现象

有时,远方天体(如类星体)的光会近距离穿过某个大质量天体(如星系或星系团)的中心。这时,中间天体的引力场就会像凸透镜那样,将光源发散的光汇聚在我们附近的某处。图31下直观地表现了这种现象。

如图31所示,由一般的引力场形成的凸透镜质量不佳,不能使所有光线都汇聚于一点,因而只能产生失焦和扭曲的图像。实际上,引力透镜可以为同一个天体形成多个像。1979年,我们首次发现了这种**强透镜成像**的实例——SBS 0957 + 561,而且它依然是令人印象最深的例子:一个红移 $z = 0.355$ 的星系给一颗类星体生成了两个相距6角秒的像。发现 SBS 0957 + 561 后,人们又展开系统调查,搜寻多处成像的类星体,并找到了几百个结果。其中有些还有四个像,但它们各个像之间的距离基本上都

小于 SBS 0957 + 561 两个像的距离。

我们在第四章的"时域天文学"一节中谈到，类星体的光度在数月或数年的时段内会有十分明显的波动。如果一个类星体有多个像，那么对于构成每个像的光子，它们从类星体传到地球的时间都是特定的。将一个像的亮度记录向前推进或向后回溯，使它与另一个像的亮度记录吻合，就可以推算二者的时间差。任何形成透镜的引力场模型都能计算这样的时间延迟，它们的值与**哈勃常数** H 成正比。哈勃常数通过 $z = Hs/c$ 将红移 z 和距离 s 联系起来，因为它决定了我们与已知红移的天体的距离：距离越远，光子到我们这里的旅途就越长。路径长了，时间延迟也就越大。

类星体和星系呈现多个像的现象很罕见。而非常常见的情况是，在我们与遥远星系连线上的引力场将星系的像扭曲和放大（图32）。这种现象叫作**弱透镜成像**。此时，引力场就像是给巨型望远镜里装了一块打磨欠佳的主透镜。经过弱透镜成像的

图32　星系团阿贝尔2218的弱透镜效应。它的引力场为背后的星系成像，将它们沿垂直场的方向拉伸为弧形

星系可能因此变亮,相比红移相同的一般星系,我们更容易细致地研究它们。

引力透镜成像的特点是质量差,但这甚至也能被天文学家加以利用。弱透镜成像的图像会向垂直引力场的方向延展。因此,如果我们通过引力透镜观察大量本身是圆形的星系,它们就会呈现为椭圆形。椭圆的短轴平行于引力场在天空的投影,椭圆的长短轴之比能表征场强。不过,我们很难真的用这种方法测量引力场的强度,因为星系的图像本来就是椭圆形。然而,引力场对像的扭曲效果,往往会将天空中距离接近的椭圆星系的像连成一排。精确地测量这种效应是当前天文学研究的一项重要方法。

微引力透镜

前景恒星经过背景恒星面前时,前者的引力场可能会对后者产生强烈的聚焦效果。通常而言,背景恒星的各个像会挨得很近,当前的望远镜无法——分辨,但我们能根据背景恒星亮度判断发生了聚焦,因为引力场将它的光汇聚在了地球上。这种形成多个不可分辨的像的现象叫作**微引力透镜**,是我们探查银河系的一项重要手段。通常前景恒星都太过暗淡,无法探测,所以我们只能看到一颗恒星突然变亮的景象。

1990年代末以来,我们每晚都监测着上亿颗恒星的亮度,发现了几千次微引力透镜事件。这些数据很难解释,因为很多恒星的光度本就起伏不定。区分光度变化和微引力透镜的方法有两种:(1)前者通常伴有颜色变化,而后者没有;(2)历史上发生过微引力透镜成像的恒星,再次发生的概率可以忽略不计,而光度变化的恒星很可能经常变化。

微引力透镜之所以重要,是因为它能让我们探测那些太过昏暗而无法直接观测的天体的引力场。实际上,任意恒星发生微引力透镜成像的概率,只取决于我们与恒星连线上的透镜天体的质量密度,而与其中某一个的质量无关。所以,不论中间的物质是质量为 $1\,000\,M_\odot$ 的黑洞、质量为 $1\,M_\odot$ 的恒星还是质量为 $10^{-3}\,M_\odot$ 的大行星,给定恒星在今晚成像的概率可能都是 10^{-6} 左右。它们的区别在于成像的持续时间,时长正比于透镜天体的质量。所以,如果引发成像的是黑洞,每次事件的持续时间会比大行星的长几百万倍,但每年发生的次数则只有后者的几百万分之一。因此,监测恒星的亮度可以确定透镜天体的密度和它们的典型质量。如果透镜天体质量非常大,我们将有幸探测到一次单独的事件;如果非常小,每次事件将转瞬即逝,不够我们在此期间观测到亮度的涨落,我们也就无法将成像事件从噪声信号中区分出来。不过这种观测可以分辨的质量范围很大,约在 $100\,M_\odot$ 到 $10^{-3}\,M_\odot$ 之间。因此,对于银河系那些质量极小的恒星和自由游荡的行星,可以用微引力透镜确定它们的空间密度上限。它还可以用来探测其他方式探测不到的行星系。

如果一颗受到日常监控的恒星发生亮度变化,暗示发生了微引力透镜成像,我们就会根据数据建立透镜天体的模型。如果模型认为背景恒星就要经过透镜中心附近(即不可见恒星所在之处),那么全球的观测者,包括许多天文观测爱好者都会收到通知。因为此时需要24小时监测那颗恒星,这不是一两个观测站就能完成的任务。在恒星非常靠近透镜中心的那几个小时,行星对透镜天体引力场的影响会使测得的亮度发生相当大的改变，从而暴露自己（图33）。

太阳对光的偏转

太阳的引力场构成了一块透镜，我们也身在其中。将我们到太阳的距离代入（6.1），可以得到一个非常接近1的折射率（$n-1 \simeq 5\times 10^{-9}$），所以射向我们的光线只会偏转极小的角度，除非它们非常靠近太阳表面，那里的 $n-1$ 要大100倍左右。恒星的光被太阳的引力场偏转，改变了恒星在天空中的位置。原则上，我们可以比对不同时期的星图找出这样的偏转。

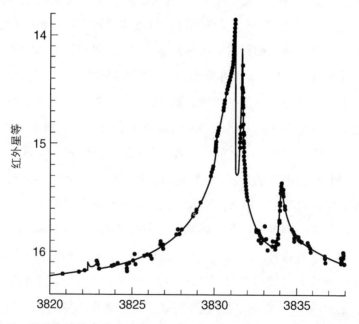

图33 微引力透镜事件OGLE-2006-BLG-109。图中画出了12家观测台测得的恒星亮度随时间的变化，单位为天。两颗行星的引力场使亮度迅速产生极大的涨落。由这些数据可以得到恒星的质量为0.51 $M_\odot$，而行星质量分别为231 $M_\oplus$ 和86 $M_\oplus$，与木星和土星相当

爱因斯坦的引力论刚刚提出、尚未验证时，靠近太阳的恒星只能在日食期间看到。此外，所需的测量精度也是一大挑战，因为即便是太阳边缘处的恒星，预计的位置移动也只有1.75角秒。而且在狭小的视场中，所有恒星的移动程度都相差不多——测量恒星的绝对位置比测量相邻恒星夹角要困难许多。尽管如此，在1919年日食期间，亚瑟·爱丁顿带领的团队还是测出了恒星的移动，结果与爱因斯坦的预计一致。

在太空中可以观测非常靠近太阳的恒星，但太空望远镜不能这么做，因为只要太阳的任何一部分进入它们的视场，其中的精度探测器就会烧坏。2013年12月发射的盖亚卫星能够以极高的精度（≥ 0.000 01角秒）测量恒星的位置，因此必须对整个天空计入爱因斯坦的移动。实际上，分析中还需要考虑行星对光的偏转。

夏皮罗延迟

1960年代起，我们能够接收行星反射的雷达波。这么做的目的，是测量脉冲回到地球的时间，并将它与使用了广义相对论的太阳系模型所预测的时间相比对。这类早期实验有两个问题：（1）雷达波不会在行星表面立即反射回来，而是会滞留一段时间，时长取决于行星的形状；（2）雷达波在行星之间并不完全以光速传播，因为其中包含稀薄的等离子体，会使折射率偏离1。这两个问题也有解决办法，可以用飞船代替行星，并设为收到信号后经过给定的延迟时间再返回信号。至于当中等离子体的影响，可以比较不同频率信号的结果来消除，因为等离子体导致的延迟与频率有关。

这些实验能直接探测太阳系内的引力场，先前我们预计它至少是卡尔·史瓦西的爱因斯坦方程解的微扰形式。结果表明，实际与预计的差异小于千分之几。

脉冲星和引力波

中子星中的许多乃至于全部都带有磁场，它们随中子星自转而扫过周围的空间。旋转的磁场会使射电波不断扫过太空，就像灯塔中旋转的信号灯扫过海面那样。波束定期经过地球，这便是**脉冲星**极具特征的射电信号。

中子星的自转非常有规律，因为在它附近很难有什么东西能对它施加明显的扭矩。由于中子星的旋转非常稳定，比较接收脉冲的时间与计算得出的发射时间，就可以对它做出精密的测量。从这个角度看，最有意思的天体当属赫尔斯-泰勒脉冲星（第90页）。两星之间的距离在75万到315万千米之间变动（作为参考，太阳的半径是70万千米）。其中的一颗中子星是脉冲星，用广义相对论计算它的射电脉冲到达的时间，会得到一个非常复杂的模式，因为每次脉冲到达我们的距离都在变化，而且它会经过两颗中子星强大的引力场，因此其中的等效折射率[见(6.2)式]也在变。预测的模式与测量结果非常吻合。

任何引力论，只要满足洛伦兹变换所揭露的对称性，都预言双星会发射引力波。因为引力场的场源移动后，场也会随之而变化，靠近场源的场先变化，远离场源的场后变化。变更过程由运动的场源发出的波完成。引力波基本的辐射机制与电磁波相同，所以有效辐射的关键是场源（天线）不能比波长小太多（参见第二章的"气体的发射"）。对于双星的情况，这指的是双星的运动速度不能比光速小太多。所以两颗几乎相撞的中子星会辐射出高频的引力波,在几个轨道周期的时间内(不到1秒)损失大量能量;而相距1天文单位左右、周期为1年的双星则是效率极低的引力波辐射体。赫尔斯-泰勒脉冲星是我们已知最好的辐射体之一。然而它算不上出色:损失能量的时标是0.3兆年左右,或者说3.4亿个轨道周期。尽管如此,由于脉冲有精确的测量,我们可以得到引力辐射导致的周期变化,结果与爱因斯坦的理论预计非常吻合。

  本书写作时,我们仍未探测到引力波[①]。这是因为,如果基本要求是大质量天体以接近光速的速度运动,那么我们就很难建造有效的天线。理想的探测器是让光来回通过两条相互垂直、长5千米的真空隧道。其中一条隧道里往返的光会与另一条的光产生干涉图案。如果有引力波经过这套装置,它会改变隧道内的等效折射率 β [见(6.2)式],使干涉图案发生移动。任何天体源的预期效果都非常小,但几年内我们应该就能观测到引力波。这无疑是实验物理有史以来最难完成的壮举。

[①] 2016年2月11日,激光干涉引力波天文台(LIGO)正式宣布发现引力波。

第七章

星　系

当我们在漆黑的夜晚仰望星空时,头上那点点的星光绝大多数是银河系的恒星(其中最亮的两三颗是行星)。在非常暗的地方,你或许能分辨出仙女星云的痕迹;如果望向极南方,还能看到大小麦哲伦星云。相比之下,目前最先进的望远镜所探测的光源大部分是星系。银河系充满了恒星,而宇宙则似乎充满了星系。

星系的形态

大体上看,星系包含大量点物质,它们在共同产生的引力场中自由运动。其中一部分是恒星,但大部分可能是某一类未知的基本粒子,它们组成了**暗物质**:我们看不见这种物质,但能通过引力探知它们。尽管恒星和暗物质的质量差异巨大,但它们的粒子遵循同样的运动过程,在共同的引力场中以非相对论的方式运动,天体物理学因而变得更加简单。然而,它们的典型轨道并不相同:暗物质粒子的轨道通常能量更高,使得它们远离星系中心。我们还认为，相比它们，恒星的轨道更加贴近星系的赤道面，其中有许多限定在赤道面的薄盘中。

不同星系中，恒星相比暗物质的质量占比差异非常大。在质量最小的一类星系——**矮椭球星系**中，恒星占有的质量不足1%。我们所在的银河系属于恒星质量占比最大的一类，比例大约是5%。不论考察哪一类星系，暗物质都占据了绝大部分质量。然而，在银河系这样的星系中，中心附近几千秒差距范围内，质量占据优势的是恒星，在这之外才由暗物质主导。矮椭球星系则不同，它在任何半径都以暗物质为主。

图34的**星系光度函数**是星系数量的一个基本性质。它展示了每单位对数光度 $\log L$ 内星系的数量。可以看到，昏暗的星系数量很多，明亮的星系则相当少。在图34左侧即低光度一侧，光度函数图像先平缓下降，在特征光度 L^* 处突然急转直下。这个 L^* 叫作**谢克特光度**，正巧与银河系的光度非常接近。

由这张星系光度函数图可以提出如下问题：如果从宇宙的所有恒星中随机选取一颗，那么它所在星系的平均光度是多少？答案是非常接近 L^*，所以它与银河系光度接近或许并不是巧合。可以想见，宇宙中有无数文明也提出并研究过这个问题，他们中的大多数也被所在星系的某颗恒星温暖地照耀着，就像我们的太阳一样。

分解星系

设想将星系分成几个**组成部分**，可以方便我们展开研究。银河系最显眼的部分是恒星盘，太阳就位居其中（图35）。我们发现，恒星盘的表面密度随半径向外大致呈指数形式下降。

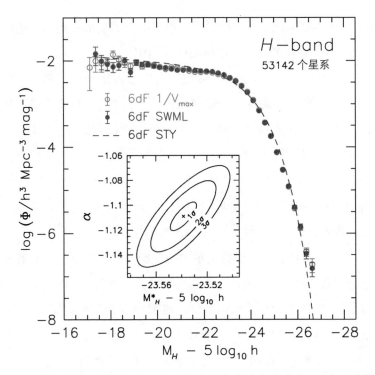

图 34　星系光度函数,描绘了星系的空间密度随光度的变化,使用对数坐标

恒星盘内部通常包含气体盘,与银河系的气体盘类似。对此我们在第二章的"气体盘"一节已经做过讨论。气体盘在半径上通常比恒星盘更大,但也薄得多。具有明显气体盘的星系,在它的气体盘和恒星盘中一般也有旋臂。若气体盘不明显,通常便没有旋臂。拥有明显的恒星盘但没有明显气体盘的星系叫作**透镜状**星系,也叫 S0 星系。我们的银河系是**旋涡**星系。

银河系内部大约 3 千秒差距的球体是**核球**(图36)。顾名思义,这个球状的部分不像银盘那样平摊在赤道面上。银河系的

图35　由5亿颗恒星构成的银河系结构图。可以看到明显的遮挡尘埃云

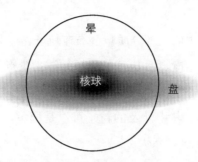

图36　盘、核球和晕的简单示意图

核球不是轴对称的（银盘近乎轴对称），而是短棒状。它的长度是宽度的三倍左右，长轴在银道面内。这根短棒像搅拌棒一样在银河系中心旋转，其中的恒星沿着偏心率很高的轨道高速运动。短棒在附近的银盘中带起螺旋波，但它们在盘内旋转的速度比短棒慢得多。

在类似银河系的星系中，核球常常呈现为棒状。但不是所有核球都如此，也不是所有旋涡星系都有核球。例如，**本星系群**中第三亮的星系（银河系是第二亮星系）——**三角座星云**（M33）就没有核球。为什么存在M33这种没有核球的星系，这个问题让宇宙学家非常困惑。

在**盘星系**中，核球从属于恒星盘（图35）；而在**椭球星系**中则相反，核球盖过了恒星盘，只有用极其细致的定量分析才有可能发现恒星盘。椭球星系一般呈轴对称形态，但也不全是。椭球星系中，恒星的运动秩序既不如恒星盘（在恒星盘中，绕对称轴附近位置旋转是绝对的主流形式），也比不过银河系等星系中的核球。

恒星动力学

由于星系中的大部分质量分布在粒子、恒星和暗物质当中，而它们又很少相撞，所以，我们需要知道大量点状物质在相互之间的引力吸引下会如何运动——这是天体物理学的一个分支，叫**恒星动力学**。其中涉及的粒子可以是恒星，也可以是暗物质粒子，处理起来区别不大。

真的可以将恒星当作点状物质对待吗？对于这个问题，答案通常毫无疑问：可以。大约30亿年后，银河系会与距离我们最近的大质量邻居——仙女星云相撞并合并。随后的几亿年间，双方会有10^{11}颗恒星相互穿行，仿佛一场极为壮观的阅兵仪式，而预计发生实际碰撞的次数却不到一次！其实，将恒星当作点粒子而可能出现问题的场景只有一个，就是极其靠近星系中心黑洞的地方。但即便是那里，也只有巨型恒星会发生实际碰撞，而且这种接触可能只会剥去恒星膨胀的大气，并不会夺走它的大部分质量。

太阳距离银河系中心大约8.3千秒差距，以近乎圆形的轨道运动。将太阳拉向银河系中心、使之保持轨道运动的引力，是构成银河系10^{11}颗恒星以及无数暗物质粒子对它的引力之和。太阳附近物质贡献的分量可以忽略不计。这与固体和液体中的情况截然不同：作用在原子上的力完全由最靠近它的少量物质主导,因为原子内的力随距离衰减的幅度,远超恒星间引力的衰减幅度。由于作用在太阳上的力主要来自大量遥远的天体,那么即使太阳向任意方向移动1秒差距,它受到的力也不会有多少变化,在接下来的几百万年里同样如此。所以,这个力是关于位置与时间的非常平滑的函数。因此,我们可以非常精确地计算太阳等恒星的轨道,方法是将每个粒子的质量均匀分摊到几个粒子的间距内,然后计算这种连续质量分布下的引力场。对于任意恒星系,这是我们做的第一步处理。之后,我们再研究这种平滑引力场中轨道的性质。我们把能够使用这种方法的恒星系统称为**无碰撞**系统。

我们可以将恒星在某个时刻的位置和速度设为 $\mathbf{x}$ 和 $\mathbf{v}$,由此确定它的轨道。如果每一对($\mathbf{x}, \mathbf{v}$)确定不同的轨道,那么轨道空间就是六维的,因为位置和速度都位于三维空间。但显然,不同的($\mathbf{x}, \mathbf{v}$)不一定会产生不同的轨道,因为一条轨道上不同时刻的($\mathbf{x}, \mathbf{v}$)当然会产生同一条轨道。

在典型的星系引力场中对轨道积分,可以得到轨道空间是三维的结果。也就是说,一条轨道可以用专门的三个数表示。它们叫作**运动常数**,因为它们的值不会沿轨道变化,恒星动力学的基本任务就是研究如何由($\mathbf{x}, \mathbf{v}$)计算适当的运动常数 J_i。也就是说,需要一种算法来计算三个函数 $J_i(\mathbf{x}, \mathbf{v})$。

星系的动力学状态可以约化为虚空间中恒星和暗物质粒子的密度分布。其中各点的笛卡尔坐标是三个数 J_i。这个空间叫作**作用量空间**。即使是银河系内恒星与暗物质粒子在作用量空间中的密度分布,我们的研究也尚有欠缺,系外星系就更是不甚完善了。

星系、气体与晶体

星系就像一升气体或一颗钻石,由大量相互作用的粒子组成。统计物理从热平衡的概念入手,对一瓶气体或一块晶体给出了相当完整的解释。所谓热平衡,就是不受干扰的系统在很长时间后达到的状态。我们根据**最大熵原理**可知系统在热平衡时的粒子分布。以气体为例,由最大熵原理可以算出以某个速度运动的分子有多少(即**麦克斯韦分布**),以及给定能量和体积下,气体施加的压强是多少。对热平衡状态施以扰动,可得到系统的**输运系数**,如声速、导热率、黏性等。

可惜的是,这条分析链的第一步就不适用于星系,因为星系没有最大熵状态,因此无法达到热平衡。

熵代表了无序。最大熵原理说白了就是:在给定能量、体积以及其他对于粒子分布的限制下,系统热平衡时会达到最无序的状态。星系、恒星这样的自引力天体总可以增大自己的熵。它们不断地将质量向内移动,增大中心附近的引力场强度,然后将压缩产生的热量释放出去,转移给外围的粒子。这些能量会增大粒子到中心的活动距离,因而增大它们的无序程度。我们在第三章的"主序之后"一节看到,在恒星生命末期,它的核心区不断收缩,包层不断膨胀。这是因为恒星在通过上面描述的过程增大自己的熵。

平衡态动力学模型

因为星系无法达到热平衡,所以我们的主要问题是推导恒星和暗物质粒子的基本分布。一旦我们知道了分布,就能算出输运系数。不过,要有了整体的构型才能施加微扰,但我们不知道用什么原理得到它。一种替代的方法是根据宇宙学来模拟星系形成过程,另一种方法是用动力学模型去拟合数据。

宇宙学模拟本身不会带来有用的预测,因为恒星和星系的形成过程极为复杂,而我们缺乏模拟的资源。因此,一切模拟都依赖于数学公式,我们希望它尽可能得出被模型忽略的物理结果。公式中的各个参数必须依照观测调整。所以,如果想构建星系模型,应当直接使用观测数据,而不是费心去模拟。

椭球星系 椭球星系最容易建模,它们也有了大量的动力学模型。从这些模型中得出的一大结论是,这类星系大多接近轴对称,并因为自转而变得扁平。然而,质量最大的椭球星系自转得非常慢,呈现三轴形态,就像李子的核。它们之所以形成自转慢的三轴形态,或许是因为它们诞生于两个质量相当、缺少气体的星系合并。

由椭球星系和透镜状星系的模型得出的另一大结论是,星系越亮,所含重元素就越多,暗物质的总质量占比就越高。大质量星系中重元素富集程度会提高,或许是因为大质量星系的引力势阱太深,超新星更难将核反应的产物送出去。暗物质占比增加,或许是因为大质量星系的恒星密度通常比小质量星系的更低。

椭球星系和透镜状星系最中心的模型特别有意思,因为它们或许能让我们探测到中心的**特大质量黑洞**。关键思路是,在半径 $r_{影响}$ 内,黑洞对引力场的贡献与所有恒星相当,那么恒星的随机速度必须增长大约 $1/\sqrt{r}$。要确保探测到黑洞,需要测量恒星密度与极靠近中心处的恒星的随机速度。哈勃太空望远镜采集的这类数据可以说是无价之宝。

我们证明,黑洞的推测质量与恒星在远大于$r_{影响}$处的随机速度密切相关。这个发现表明,黑洞的增长与星系恒星群的增长有因果关系。这个结果出人意料,因为恒星群的质量和延伸范围都比黑洞大多了。然而,类星体(即高速吸积冷气体的黑洞)的空间密度正好在红移z约等于2时达到最大,此时宇宙中的恒星生成率也最高。因为恒星生成于冷气体,黑洞以及它所在星系的恒星群的生长速率很可能都仰仗于冷气体,所以它们当前的质量密切相关也在情理之中。

旋涡星系 旋涡星系中,得到最广泛建模的是我们的银河系。我们已经为它的薄盘与厚盘建立了理想的平衡态动力学模型,由它们的垂直结构可以推出,使太阳维持轨道运动的力中,56%由暗物质产生,所以只有44%由恒星产生。银盘的质量与只包含恒星和气体、没有暗物质时的一致。

缓慢漂移

在银河系的整个生命周期内,没有一颗恒星会一直留在同一条轨道上,因为我们在计算运动常数J_i时假设的平滑、与时间无关的引力场是一种理想情况。真实的引力场会出于各种原因在理想值上下浮动。第一,理想情况没有体现银盘的旋涡结构。第二,银盘中包含大量分子云,它们在盘中以各种方式变形、移动、消散。这些物质的引力场在理想情况下被忽略了。第三,任何星系都不是孤立的,其他星系(其中许多规模不大)时常会并入银河系这样的大星系,它们会在星系里穿行很长时间才消散。在理想情况中,我们忽略了这些运动的大质量团块。最后,任何体积的点物质数量都有"泊松涨落"：如果粒子密度满足在体积 V 中预计平均含有 N 个粒子，那么该体积内粒子数量随时间的实际涨落大约为 $\sqrt{N}$。由于体积 V 产生的力正比于它所包含的物质数量，因此引力的涨落是自身的 $\sqrt{N}/N = 1/\sqrt{N}$。

　　设想一颗恒星或暗物质粒子在理想的平滑引力场中做轨道运动，我们必须假定它受到一个小的随机场扰动。由于这个随机场影响，原本在轨道 **J** 的恒星或暗物质粒子，就有概率在时间 t 变换到轨道 **J'**。这与水面上的花粉所做的布朗运动非常相似：它们似乎在随机地跳动，所以在短时间 t 内有概率从位置 **x** 移动到临近的位置 **x'**。每颗花粉都在这样随机运动着，整体结果就是花粉密度在空间中**扩散**：如果花粉最初都集中在位置 **x**，那么一段时间后，它们会因为扩散从 **x** 散布到整片空间。同样，恒星和暗物质粒子也会在作用量空间中扩散。

　　对于我们所在的恒星盘中的恒星来说，扩散非常重要，因为恒星在作用量空间中一处特别限定的区域内形成轨迹，这条线与它在恒星盘对称面的圆周轨道有关。恒星若从这条线向外扩散，轨道偏心率就会增大，相比对称面也更倾斜。于是，恒星的随机速度会增大。由于气体分子的随机速度与热量有关，银盘因此"变热"。这个比喻不那么恰当，因为并没有什么东西在为银盘输送能量，使它变热。它是自发、自动地加热，从引力势能中抽取随机速度增大所需的能量。引起银盘加热的引力场涨落主要由旋涡结构（下一节讨论）产生，但大的分子云也有重要贡献。我们尚不清楚坠入的矮星系是否也发挥了重大作用。

　　球状星团（见第三章的"球状星团"）非常像小型星系。由于恒星密度发生上面提到的泊松涨落 $\sqrt{N}$，它们的核球收缩，包层膨胀。在星系中，要等到泊松涨落引起可观测的收缩，所需时间比宇宙的年龄还要长。

星团的破坏 大部分恒星诞生于小星团，这些星团的质量不超过 $1\,000\,M_\odot$。其中的泊松涨落十分显著，恒星在相对较短的时间内就能完成能量分配。这种星团中的恒星只需要少量动能就能完全脱离，而恒星在任意一次能量交换中，获得能量的一方总有机会得到足以逃逸的能量。脱离的恒星不会再返回并失去能量。由于星团与逃逸恒星的总能量守恒，失去逃逸者的能量必然使留下的星团能量减少。这个过程本质上和蒸发冷却一样（所以我们湿透之后站在风中会觉得冷）。

随着星团因为恒星"蒸发"而收缩，泊松涨落的作用也变得越来越大，蒸发率并没有因为恒星间的平均束缚力增强而降低。最终，星团收缩成一对双星；这个过程释放的能量足以让其他恒星永远脱离。

上面描述了一个孤立的小型星团在很长时间内的变化过程。其实，星团并非彼此隔绝，它们会在星系中穿行。我们将在第131页看到，它们会在星系的引力场中逐渐瓦解。实际上，太阳等如今不属于任何星团的恒星，可能就是从某个星团中逃逸出来的。

旋涡结构

银河系这样的星系包含旋臂。我们在第四章看到，恒星与黑洞外围吸积盘的原理就是角动量向外传递，这个过程加热了吸积盘。同样，角动量向外传递对银盘也非常重要。我们在第四章中看到，气态吸积盘的角动量主要由磁场传递。在恒星盘中，磁场的动能由引力场完成，因为恒星之间只受引力作用。旋涡结构提供了角动量向外传递所需的引力场。

除了将角动量带到恒星盘外，旋臂还会频繁冲击星际气体形成激波，使之变得更致密，其中一部分会坍缩形成新恒星。因此，旋臂结构最容易在年轻恒星分布的地方找到，尤其是那些大质量的明亮恒星，因为所有大质量恒星都很年轻。它们一生短暂，不会离开激波中的出生地太远，所以它们能标记激波的狭窄轨迹。引发激波的引力场大多由大量更老、质量更小的恒星产生。它们的分布构成了更平更宽的旋臂，在用红外线拍摄旋涡星系时最为显眼。

旋臂就像声波在空气中传播那样，是在恒星盘中传播的高密度恒星波。它们也和声波一样携带能量，这些能量最终会从波这种有序的形式转变为恒星随机运动的动能。也就是说，旋臂加热了恒星盘。声波会加热所到之处的空气，同理，旋臂会加热特定半径处的恒星盘。在那里，恒星会与波共振。波与特定位置的粒子交换能量是无碰撞系统的一个特征，也是电子等离子体运动的关键。我们对波与粒子的相互作用依然认识不足，对它们在旋涡结构和星系形态演变中起到的具体作用也尚有争议。

核球的成因

用N体构成的短棒模拟自引力恒星盘，很好地复现了核球内数千恒星沿视线的速度。模拟结果表明，形成短棒需要两步：首先是形成一段很扁的条，稍后扁条发生紊乱，在垂直方向变宽。由此得到的数据与如下假设一致，即核球与恒星盘中所有的恒星都形成于占据了赤道面的薄气体层。这解释了银河系绝大部分恒星的来源：恒星晕中的恒星可能并非形成于银道面，但银河系中只有不到1%的恒星属于恒星晕。

对暗晕中的短棒做N体模拟，会发现短棒将角动量转移给暗晕，导致自身的旋转速率变慢，同时强度提高。许多证据表明银河系的短棒有很快的旋转速率，这与N体模型的结果一致，前提是有一股很强的气流穿过银盘进入棒中。实际上，气体向内移动，是因为它在经过旋臂时损失了角动量。当气体来到短棒的共转半径时，就开始从短棒旋转的引力场中获取角动量，因此气体向内移动的速度减慢，密度增大——这就是**巨分子环**的来源。这是一片富含气体的区域，半径约为5千秒差距，环绕在短棒外围，主导了银河系恒星的形成。从巨分子环中脱离并进入棒中的气体会在激波中迅速失去角动量，这在数值模拟和系外星系的尘埃带中都能看到。气体迅速冲入棒中后，会在一个半径约为0.2千秒差距的核盘中累积，这个区域叫作**中心分子区**。核盘内外不远处有无数超新星残骸，由此可以证明，气体在这里进行过激烈的造星运动。

同类相食

宇宙包含着许多暗物质晕，引力在其中扼制宇宙膨胀。本星系群就是这样的晕，银河系是其中三个大型星系之一。因为矮星系的形成数量远多于巨星系，所以任何晕中的星系都以矮星系居多。

每个晕的边缘都有许多矮星系，它们处在从晕的中心向外扩散和落回晕中心的平衡之中。它们仍受晕的引力场作用，运动轨道高度偏心，此时正好在距离晕中心最远处。几十亿年之后,它们将非常靠近晕的中心。这种经历会对它们带来什么影响呢?

星系从$r_{远}$处落向晕的过程中,会吸引路径附近的暗物质粒子。要理解这个行为的后果,最简单的方法是将暗物质粒子想象为流体。矮行星经过时,流体受到沿运动方向的冲量,从而向运动路径汇聚(图37)。然而水流沿这条线移动、汇聚都需要时间。所以密度变大的暗物质区域会在矮星系后方。于是,这片区域的引力会将矮星系往回拉:引力阻滞了矮星系的运动,使它像是受到了摩擦力作用。实际上,这种现象就叫作**动力学摩擦**。

使暗物质形成致密区域的冲量与矮星系的质量成正比,所以致密区的质量也与矮星系的质量成正比。矮星系受到的拖拽力正比于矮星系质量与致密区质量的乘积,因此与矮星系质量的**平方**成正比,矮星系的减速度正比于自己的质量。所以,落入的星系质量越大,动力学摩擦就会越快地改变这个星系。

如果矮星系没有受到动力学摩擦影响,它就会从晕的中心附近穿过,运动到半径与起始位置几乎相等的位置。它不会正

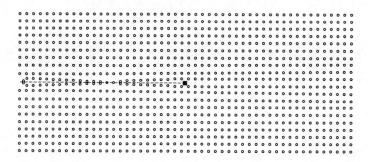

图37 大质量天体在身后形成尾迹。中心的黑色方块代表大质量天体,它在从左向右移动。网格点处原本静止的粒子纷纷向它移动。显然,大质量天体背后的轨迹上,有一块区域的粒子密度变大了

好回到$r_{远}$的地方,因为其他落入的物质增大了$r_{远}$范围内的质量,使它没有足够的能量到达$r_{远}$处。如果矮星系受到很大的动力学摩擦,那它还没到$r_{远}$就会停止运动。动力学摩擦使得每次折返的半径都比上一次更小,最终矮星系会停留在晕的中心附近。因此,这个晕,或者说处在晕中心的星系,就把矮星系"吃"掉了。

如果事情如上面所述,那么质量为m的矮星系落入质量为$M(r)$的晕后,存留的时间数量级为

$$t_{吞} = \frac{M(r_{远})}{m} T,$$

其中T是在半径$r_{远}$处做圆周运动的周期。如果合并的两个星系质量相近(比如银河系与仙女星云M31未来的情形),即$m \simeq M(r)$,那么套用公式可得$t_{吞} = T$,它被称为初始轨道周期。N体模拟结果表明,这个推算是正确的。

当前最靠近银河系的矮星系是人马矮星系,距离银河系中心大约13千秒差距,位于与太阳相对的一侧。这个半径下的圆周轨道周期T约为5亿年,接近当前宇宙年龄的1/25。因此,要让$t_{吞}$比宇宙当前年龄更小,矮星系的质量就必须大于$10^{10} M_\odot$,因为$M(20\ \text{kpc}) \simeq 2.5 \times 10^{11}$。虽然矮星系质量还无法精准确定,但它肯定远小于$10^{10} M_\odot$,因此可以认为它不会很快被吃掉。但结果并非如此!因为我们低估了银河系的胃口。

随潮而散

一个具有一定体积的物体在引力场中运动时,会沿自己的中心与引力源中心的连线拉伸。原因是,它表面上最靠近引力

源的那部分物质所感受到的引力,大于远离引力那端的物质所感受到的力(图38)。所以,这两块物质会以不同的加速度运动。但它们又在同一个物体上,于是只能各退一步,以同样的加速度运动。这个加速度对靠近引力中心那端的物质来说太小,对另一端的物质又太大了。对此,物体沿二者中心连线方向拉伸。于是,它的引力场会将靠近引力中心的物质拉回,将另一边的物质推远。由于这里讨论的就是月球在海洋表面引发潮汐的物理过程,我们便说这样的天体被它所环绕天体的**潮汐场**拉伸了。

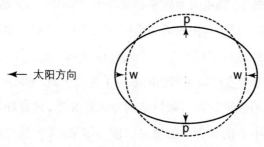

图38 没有月球的潮汐:在地球距离太阳最近和最远的点上,海平面(实线)高于平均水平线(虚线),因此水压太小,无法平衡地球的引力。此处一小体积的水受到的净力w朝下。在左侧,净力抵消了一部分太阳的引力,而在右侧,它与太阳的引力叠加。二者对太阳引力的增减,确保了两处的海洋环境以相同的角速度绕太阳运动

矮星系向内移动的过程中,它的暗物质晕也会拉伸,乃至于其中的粒子完全脱离矮星系束缚,独自在银河系的引力场中运动。在靠近引力中心那端离开的粒子会进入角动量比矮星系更小的轨道,因此会超过它;另一端离开的粒子则会进入角动量更大的轨道,因此会落在它身后。于是就形成了两道**潮汐尾迹**,矮星系的质量也越来越小(图39)。

矮星系质量变小后,引力场随之变弱,粒子从矮星系自由脱

图39 模拟潮汐尾迹的形成。星团不断地碎裂,曲线画出了星团中心的轨道

离进入独立轨道的边界点也逐渐向矮星系中心移动。质量损失进入恶性循环,矮星系的质量越来越小,潮汐尾迹越来越长。直到某一阶段,粒子自由脱离的点位直逼中心,大量恒星以及暗物质粒子已经流入了尾迹之中。人马矮星系就进入了这个阶段,它的尾迹带着它的恒星将银河系绕了至少一圈,也可能是两圈。

我们测量了数百万颗恒星的颜色和亮度,这样便可以研究以太阳为中心的球壳内的恒星密度。结果是,恒星密度图像充满了突脊和密集的团块。这些突脊和团块包含了恒星晕中至少一半的恒星。所以,恒星晕有可能全部由矮星系和球状星团中剥离的潮汐流构成。这些天体初次进入银河系后,质量大概很大,能感受到强烈的动力学摩擦。但进行轨道运动一段时间后,潮汐使它们不断损失暗物质晕,直到它们的质量低于某个阈值,不再感受到显著的动力学摩擦。然而,潮汐剥离依然不断发生,它们最终在进入银河系中心之前便被完全吞噬。

化学演化

银河系形成以来,几乎所有比锂更重的元素都已生成,它们形成的同时,恒星也在继续演化。所以,不同年龄的恒星包含了星际介质中重元素含量演变的"化石"记录。而且，越古老的恒星往往有越大的随机速度。所以，恒星的化学组分与它的运动学性质有关。

我们很难确定一颗恒星的年龄，因为这需要精确测量恒星的质量、光度和化学组分，其中质量尤其难以确定。所以，天文学家喜欢用比较容易测量的化学组分作为年龄的替代。

星际介质包含丰富的重元素。对此，超新星的贡献最大。在第三章的"恒星爆发"一节中，我们看到有两类不同的超新星：一类是标志了大质量恒星（$M > 8\,M_\odot$）死亡的核坍缩超新星，在白矮星吸积太多伴星物质后出现。因为大质量恒星的寿命很短，恒星形成后很快（约1 000万年）就会出现一系列核坍缩超新星爆发；而演化为白矮星后会进入重要的吸积过程，可能会持续大约10亿年。所以，在银河系的前10亿年中，只有核坍缩超新星在丰富星际介质的化学组分。

我们在第三章的"恒星爆发"一节中看到，暴燃超新星产生的大部分物质是铁，而核坍缩超新星会产生很多种类的重元素。由此，在银河系的头10亿年间，星际介质中铁的含量占比，相比镁和钙等会比现在更低一些。图40是太阳附近可以确认的两组恒星群：在同样的铁元素丰度下，**α增丰**恒星群中镁和钙的丰度比**正常丰度**恒星群中的更高。由此，我们认为α增丰恒星群形成于星系的前10亿年左右。

两个恒星群中，相比于氢，铁的丰度变化范围非常大。一个可能的原因是，铁氢比低的恒星比铁氢比高的恒星更早形成。但这个结论一般而言是错的，因为在银河系早期，内部星际气体聚集为恒星的速度很可能比外部更快，因此重元素丰度在中心

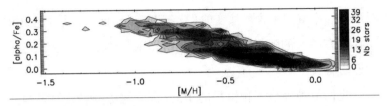

图40 前10亿年形成的恒星中,铁的丰度比镁更低,在图中的上方。横坐标是铁的丰度。等高线标示了恒星密度分布

升高得更快。于是,小半径处更快达到一定的铁氢比,那么铁氢比低的恒星要么是银河系早期在中心附近生成,要么是晚近才在大半径处生成。

这两种可能性可以由恒星的镁铁比分辨:如果镁铁比很高,那么恒星必然形成于银河系的前10亿年,那时大半径处只有铁氢比低的恒星,所以镁铁比高且铁氢比高或中等的恒星一定在相当远的区域内形成。

α增丰恒星具有很高的随机速度,相比正常丰度恒星离银道面更远。数据表明,厚盘面主要由α增丰恒星组成,薄盘面主要由正常丰度恒星组成。

考虑到恒星诞生的时间与地点、化学组分以及恒星运动学之间有紧密联系,我们很有必要建立化学和动力学演化的共同模型。在这样的模型中,恒星以一定速率在各半径处从当地的星际气体中生成,并在之后迅速增加镁和钙的丰度,又在10亿年及之后提供了铁。恒星诞生时的轨道近乎圆形,然后逐渐向偏心率更高、更倾斜的轨道漂移。各半径处的星际气体因生成恒星和超新星加热引起星系风而减少,又通过吸积星系际气体而增多。旋涡结构驱使生成恒星的气体缓慢向内移动,它们带来了重元素。这个模型的目的是复现观测到的恒星运动(它是位置的函数），以及化学组分与运动学之间的关系。这是当前研究的一个活跃领域。

大储藏室

在第二章的"气体盘"一节中，我们描述了银河系的气体盘，它是典型的旋涡星系气体盘，质量约为 6×10^9 $M_\odot$。由于银河系以 2 $M_\odot yr^{-1}$ 的速率将气体转化为恒星，这样的物质储量会在30亿年内耗尽。当前的气体盘究竟是一块更大的、形成了大约 5×10^{10} $M_\odot$ 恒星的气体盘的残余，还是恒星形成与气体吸积的缓冲区呢？

根据其他星系的观测结果，气体盘中的恒星生成速率与冷气体的表面密度成正比，所以没有吸积时，气体盘的质量会随时间以指数形式下降。由银河系当前生成恒星的速率以及当前的气体质量，很容易证明，如果没有吸积，过去100亿年的气体质量大约是 1.7×10^{11} $M_\odot$，这些质量如今几乎都变成了恒星。这个荒谬的结论表明，银河系不吸积气体的前提是错误的。

由宇宙微波背景可知当前宇宙中普通物质的平均密度。测量星系的光度得到宇宙平均光度密度，由此可知平均需要多少普通物质才能产生特定的光度——大约是 40 $M_\odot/L_\odot$。如果任意选择具有代表性的星系群，由它的光度便可以推断它应当包含多少普通物质。结果表明，它大约是星系中普通物质的十倍。所以大部分普通物质必然位于星系*之间*，而不是星系内部。

星系间的氢原子可以通过它吸收**莱曼**α光子探测，这些光子由遥远的类星体（见第四章的"类星体"）发出。因为这样的光在到达我们这里时经历了宇宙的大部分阶段，所以它几乎对宇宙每一时期的星系际空间都做了"采样"。因此，测量莱曼 α 谱线能够让我们追踪宇宙中氢的密度随时间的变化。在头 10 亿年里，氢的密度与根据宇宙微波背景推断的普通物质密度差不多。但随着时间推移，氢的密度不断下降，如今它已经不足普通物质预计平均密度的 1%。探测邻近星系之间氢原子的 21 厘米发射线，也证明了现在的星系际氢原子已经所剩无几。

对于这个发现，比较自然的解释是，宇宙中的氢被用去创造恒星和星系了。但对邻近星系的研究表明，它们包含的普通物质不足，上述解释并不成立。可以接受的解释是，消失的物质**的确**在星系际空间，但它们十分热，已经完全电离，因此不能根据氢的谱线探测。

实际上，星系际气体十分热也是很自然的事情，因为只有气体的温度超过**位力温度**，气体的压强才能对抗星系的引力。在这个温度下，原子的热运动速度与暗物质粒子的轨道速度相差无几。对银河系而言，这个温度大约是 2×10^6 开，但它在宇宙各处是变化的，在星系最为密集的地方可以达到 10^8 开。在位力温度下，气体压强能够有效地平衡引力，所以气体密度往往能反映暗物质的密度。若气体中原子的运动速度远低于暗物质粒子，那么它的内部压强就太小，无法抵抗引力。平衡态的气体必定会被限制在薄的旋转盘中。

处于位力温度的气体会发射 X 射线。密集的星系团发射的 X 射线很强，可以通过探测发现，结果表明其中存在预期的气体量（图 41）。我们认为密集星团外的 X 射线十分微弱，频率也低，无法用当前的望远镜探测。然而，我们在某些天体的紫外线光谱中发现了预期气体的踪影。

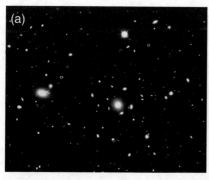
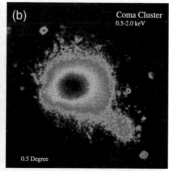

图41 后发座星系团，左边（a）是光学图像，右边（b）为X射线图像。光学图像的宽度只有0.32°，而X射线图像的宽度是2.7°，因此光学图像只显示了星系团的中心。光学图像上方沿中线靠右的明亮天体是银河系中的一颗恒星，其他天体都是星系

目前，探测气体最灵敏的方法是找到背景天体光谱中的吸收线（第137页）。检测吸收线的最佳波段取决于你要寻找的气体的温度，因为波段包含的光子应当具有正确的能量，使离子从基态进入激发态。而在高温时，只有牢牢束缚的电子还围绕在离子周围，这些离子只能被高能光子激发。

探测温度大于10^6开的气体时，理想波段是X射线波段。可惜，X射线望远镜很难建造，因为X射线往往会把反射镜的电子打出来，而不是被镜面反射。再者，给定光度下，X射线光子的发射率只有可见光光子的千分之一，所以X射线光子十分稀少，统计上的噪声是一大问题。由于这些原因，我们不可能灵敏地探测X射线的吸收线。不过，哈勃太空望远镜已经在探测紫外线光谱的吸收线。类星体的谱线表明存在五阶氧离子O^{5+}、三阶碳离子C^{3+}之类的离子。这些吸收线的速度表明，吸收发生于视线经过星系之时，但距离往往很远，通常有100千秒差距。

测得的离子在温度约为3×10^5开的气体中很常见,这个温度比我们认为的气体温度要低。对此,通常的解释是,我们看到的是填充在星系际空间和更小、更冷(约10^4开)的气体团交界处的吸收线。这是研究的一个热门领域,我们的解释可能过几年就不一样了。

决定形态的因素

星系的性质大体上由三个数决定:光度、核球与星系盘比(核盘比)、冷气体与恒星质量比。因为恒星由冷气体形成,最后一个比值决定了星系恒星群有多年轻。

年轻的恒星群包含质量大的恒星,它们寿命短、亮度高、呈蓝色(图3)。年老的恒星群只包含质量小、暗淡、呈红色的恒星。此外,年轻恒星在空间上呈团块分布,因为它们来不及扩散到整个星系中,就像刚刚倒入咖啡的奶油泡沫那样一团一团或一条一条,搅拌之后就很快不见了。年老的恒星则相反,分布得比较均匀。因此,带有年轻恒星群的星系与带有年老恒星群的星系看起来差别很大:相比恒星质量相近、带有年老恒星群的星系而言,前者有更多团块和条纹,颜色更蓝,光度更高。

核盘质量比显然影响着星系的形状,这在从侧面观察星系盘时尤为明显。它还会影响星系的紧密程度,因为特定光度的核球往往比相应的星系盘更致密。

最后,星系的结构主要由它的光度决定,因为光度与恒星质量有关,恒星质量又与恒星和暗物质粒子运动的速度有联系。因此,光度与位力温度有关。一个质量大、光度高的星系有很高的位力温度,这会使超新星更难将气体逐出星系。反过来说,超新星很容易将气体逐出质量小、光度低的星系。今天的冷气体就是明天的恒星,早早就将气体逐出星系会降低恒星与暗物质的质量比,而这被认为是光度低的星系中暗物质与恒星质量比很高的原因。

星系的形状也由所处环境塑造。密集的环境富含椭圆星系和透镜状星系,异常稀疏的环境则多见不规则矮星系。银河系等旋涡星系通常处在中等密度的区域。这些地方由直径为几兆秒差距的区块组成,其中的宇宙膨胀被引力逆转,因此每片区块现在都由大量朝彼此下落的星系构成。银河系就处在这样的区域中,属于本星系群。这种星系群中有少量星系相对当地星系际气体静止。它们是气体的汇聚之处,随气体逐渐冷却。在目前还没有完全理解的过程下——可能与超新星逐出气体盘的冷气体云(见第二章的"气体盘")的动力学相互作用有关,位力温度的气体冷却进入星系盘,补充了冷的、可形成恒星的气体。因此这些星系,包括银河系和邻近的M31、M33,依然都十分年轻。

更小的星系沿轨道环绕并穿过那些大的星系。它们无法通过吸积位力温度的气体来补充冷气体,因为它们穿过气体的速度太快了。于是,在最初储存的冷气体消耗殆尽后,它们之中形成恒星的活动也就停止了。所以伴星系距离宿主星系中心越近,我们就越不可能看到它形成恒星:位力温度气体的密度向内不断增大,所以靠近的星系在以音速穿过位力温度气体时,都会感受到强烈的"风"。这股强风(好比乘坐在敞开机舱的波音747上)会将伴星系本身的气体一扫而空(图42)。

星系团 密度最高的环境是富星系团。这些区域的大小约为几兆秒差距,其中引力很早就逆转了宇宙膨胀,因此星系的密

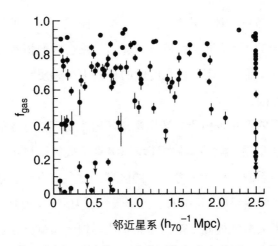

图 42 矮星系距离邻近星系越远,越有可能包含少量可以形成恒星的冷气体

度和位力温度都很高。由于位力温度气体很致密,它们发出的 X 射线也很强,即使距离星系团中心很远也能用 X 射线望远镜观测到(图41)。星系团中位力温度的气体极其炽热,因为宇宙结构的特征速度和温度会随质量增大而增大,而富星系团的质量非常大(约为 10^{15} $M_\odot$)。

富星系团的位力温度超过了超新星加热的星际气体的温度。因此自大爆炸之后,这些区域的任何普通物质都不会被吹出来,普通物质与暗物质的质量比也应当是宇宙平均比例。事实确实如此,结果在实验误差范围内。

穿越富星系团的星系无法通过冷却获得星系际气体,所以它们很难形成大量的恒星。具有冷的、能够形成恒星的气体盘的星系不断落入富星系团中,它们形成恒星的速率不断下降,直到储存的气体完全耗尽。在这一阶段,它们被称为**弱臂星系**。弱臂星系不再形成恒星后,就变成透镜状星系——它们的恒星盘是先前能够形成恒星的气体盘的化石遗迹。

通常（但并非总是如此），富星系团中心有一个质量极大的星系。这种**星团主导**星系是个特殊的庞然大物，它与本地星系际位力温度气体保持相对静止，所以能够吸积气体，应当有很高的恒星生成速率。在这类星系中，有一部分确实拥有大量年轻恒星，比如英仙座星系团中心的 NGC 1275，但大部分却并非如此，而且那些拥有年轻恒星的星系也缺少主导恒星盘。

为什么星团主导星系没有形成巨大的恒星盘，使母星系变成银河系这样的巨型旋涡星系？若是如此，星团主导星系就会成为核球，星团则是伴星系的晕。对此，天体物理学家没有完备的答案，但最终的结果显然涉及两个关键的物理因素。其一，等离子体冷却的方式会在 10^6 开左右发生根本改变。因为在这个温度下，常见的元素碳和氧会失去最后的电子。有束缚电子的离子，即使密度非常低，也能极大地增强等离子体的冷却能力。因为相比自由电子，束缚电子辐射光子的频率要**高得多**。对于给定密度的等离子体，这一现象会使冷却时间在温度从 10^6 开上升到 10^7 开时陡然提升，这妨碍了星团主导星系在外围形成冷的、能生成恒星的气体盘。

第二个关键因素是，大多数明亮的星系中心存在着特大质量黑洞（约为 $10^9\,M_\odot$）。任何位力温度的气体都会在星团主导星系中心和黑洞附近冷却，气体在那里处于最高的压强之下，密度也最高。黑洞吸积一部分气体后，会形成喷流（见第四章的"喷流"）。喷流从附近的位力温度气体中冲过，将它们重新加热。测量星团主导星系的射电频率以及 X 射线，可以明确证明这个过程。尤其是某些星团的 X 射线谱证明，那里的气体温度至少是星团主体气体温度的三分之一，但气体并没有在冷却。这个发现表明，重新加热阻碍了中心形成稳定状态，这使气体无法平稳地流入中心天体。

银河系的中心黑洞比典型的星团主导星系的中心黑洞要小得多，质量是 $4\times 10^6 M_\odot$。然而，它能够重新加热包裹它的等离子体，而且可能是定期加热。目前它似乎处在静息状态，像是一座休眠火山。这种能力解释了为什么银河系的中心不是恒星生成最激烈的区域——这项荣誉属于中心分子区。由前文对银河系核球的讨论可知，冷气体从半径约为5千秒差距的巨分子环不断输入这片半径约为0.2千秒差距的区域。对于"为什么星团不由某个巨型旋涡星系主导"这个问题，关键是解释这些星团为什么没有类似巨分子环的结构，而这或许可以追溯到等离子体在 10^6 开附近冷却方式的变化。

第八章
总览全局

我们在第六章讨论了许多可以用广义相对论解释的现象。然而，目前广义相对论对天体物理学最大的贡献是，我们可以讨论整个宇宙的几何结构和动力学，这使宇宙论不再隶属于哲学或神学，而成为物理学的一支。这里没有足够的篇幅系统地探讨宇宙论，对此，读者可以阅读《宇宙论》（"牛津通识读本"）。现在，我只概述我们对大爆炸后恒星和星系如何形成的大体理解，为前面各章介绍的物理过程提供一些背景。

宇宙论的核心要素是三种流体：无人能够解释的**暗能量**、无人能够看见的暗物质，以及宇宙微波辐射。宇宙微波辐射在红移 z 约为 3 000 之前占据主导，构成了宇宙微波背景，我们能够进行细致的研究。暗能量在较近的时期（z 约为 0.5）才成为主导，而它们之间的时期，主导是暗物质。从宇宙论的角度看，三种流体的区别是它们施加的压力。辐射施加正压力，暗物质施加的压力微乎其微，暗能量施加负压力，也就是张力。

根据广义相对论，压力和质能都是引力的来源。由于太阳内部的压力很高，它对地球的引力比不考虑压力时更大。相反，张力会生成斥力，暗能量因此而产生的斥力比它的能量密度产生的引力更大。由于暗能量目前主导着宇宙，宇宙已经被它产生的斥力撕裂。我们测量暴燃超新星（见第三章的"恒星爆发"）的红移和距离，发现了这个现象。由测量结果可以推断宇宙过去的膨胀速率。结果表明，在 z 约为 0.5 时，膨胀速率开始变大，而之前它一直在随时间变小。

大爆炸之后的头 20 万年里，宇宙由辐射主导，几乎完全均匀。因此，此时引力非常强，它使宇宙火球的扩张速率逐渐减慢。因为辐射流施加压力，所以它对膨胀做功，因此能量密度比暗物质下降得更快——暗物质由于压力微乎其微，做的功可以忽略不计。于是，在红移 z 约为 3 000 时，辐射的能量密度下降到暗物质能量密度之下。此时，能量密度原本微小的涨落开始急速增大，因为此时生成引力场的主力不再是辐射，而是暗物质，压力无法阻止某些区域变得更加致密。在这一阶段，普通物质与邻近均匀的辐射流紧紧锁定，因而没有参与暗物质形成团块的过程。之后在红移 z 约为 1 000 时，辐射的温度降至电子能被质子和 α 粒子束缚、分别形成氢原子和氦原子的程度。这些原子有效地将普通物质从辐射流中剥离，因为原子很少散射光子。此时，普通物质已经无法阻止引力将它们拉入暗物质聚集区，各种结构真正开始成型。这就是**退耦期**，我们可以测量宇宙微波背景的各种性质，对它进行详尽的研究，因为其中的光子在退耦期后再也没有受到干扰，直至传向我们。

在退耦期期间，暗物质的密度涨落只有十万分之几，所以引力花了很长时间才将涨落增大到足以使密度高的区域停止膨胀，坍缩成恒星和星系。这个过程开始于红移 z 约为15之时，首批坍缩的天体有大质量恒星等。这些恒星辐射的高能光子逐渐将氢原子和氦原子重新电离。这个过程大致在 z 约为6的时候结束。

目前，对于红移 z 位于 1 000～6 区间内的情况，我们没有太多的观测数据，但红移6之后的观测记录就很多了。此时肯定形成了少量质量非常大的星系，因为已知的明亮类星体都在 $z>7$ 之外，而且我们知道（见第四章的"类星体"）它们由大质量星系中心的大质量黑洞驱动。

虽然在 z 约为6时有少量大质量星系，但如今的恒星只有极少一部分是那时形成的。那时的大多数星系要比现在的小得多，而且富有致密冷气体。随着这些气体逐渐形成新的星系，它们也在以越来越快的速率形成恒星，直到 z 约为2时。流入星系的气体混沌无序，通常不会有组织地形成银河系现在这种扁平的气体盘。相反，它们会四处流动、互相碰撞，并在碰撞后快速形成恒星。

大质量黑洞趁着宇宙混乱不堪，疯狂吸入气体。所以星系的核球与黑洞在此时快速成长。黑洞吸积释放的能量被周边稠密的气体转化为可见光和红外线光子，使黑洞附近区域像是类星体一样明亮。大质量恒星死亡时释放的能量将附近的星际气体加热（见第二章的"气体盘"），使星系内部和附近越来越多的地方被温度大于等于位力温度的气体占据。比位力温度更热的气体向外流入星系际空间，带走了新近死亡恒星合成的大部分重元素（见第七章的"决定形态的因素"）。

从红移 z 约为2开始，气体流入星系的步伐逐渐放缓，恒星形成与黑洞吞食物质的速率也随之下降。更多的气体因为过热而无法形成恒星，也无法成为黑洞的饵料。在第七章的"决定形态的因素"一节中，我们描述了这种整体趋势对单个星系形态的影响。

以上就是宇宙的简明历史。其中涉及的物理原理大多非常复杂，我们还远远不能尽数解释当中的过程。因此，如果继续深究细节，我们很快便会触及当前知识的极限。

宇宙是一块巨大的画布，大自然在上面施展了许多技巧。我们正在快速了解这块画布以及画家的技法，但我们要学的东西依然还有很多很多。

索 引

(条目后的数字为原书页码，见本书边码)

21 cm radiation 21 厘米辐射 15, 17—18, 137

51 Pegasi 飞马座 51 71

A

absorption lines 吸收线 138—139
accretion 吸积 24, 40, 51—53, 76, 90, 127, 148
action space 作用量空间 121
aether 以太 92
alpha elements 阿尔法元素 41
Ampere, Andre-Marie 安德烈－玛丽·安培 3
Andromeda nebula 仙女星云 115, 119, 131, 141
angular momentum 角动量 5—6, 50, 66, 73, 77, 128, 132
argon 氩 45
asteroid 小行星 77
astronomical unit 天文单位 8
atomic 原子
    hydrogen 氢 19
    structure 结构 24

B

Balbus, Steve 史蒂芬·拜尔巴斯 61
Balmer 巴耳末, 见 photon

bar 短棒 128
binary 双星, 见 star
black-body 黑体, 见 radiation
black hole 黑洞 39, 53—54, 59—64, 68, 88—90, 110, 119, 123, 143, 147—148
    fluctuation timescales 涨落时标 69
    formation 形成 26
    galactic centre 银心 55
    supermassive 特大质量 55
boundary layer 边界层 58
Boyle's law 波义耳定律 21
breakup spin 裂解自转 59
bromine 溴 38
Brownian motion 布朗运动 125
bulge 核球 118
    bulge-to-disc ratio 核盘比 139
    origin 来源 128

C

calcium 钙 41, 134
calculus 微积分 2
cannibalism 同类相食 129
carbon 碳 23, 41, 139, 143
    burning 燃烧 26, 40—41
    core 核心区 29
    ignition 点火 30, 40
    monoxide 一氧化碳 14, 16, 19
central molecular zone 中心分子区 129, 144
Chandrasekhar, Subrahmanyan 苏布拉马尼扬·钱德拉塞卡 7
chemical evolution 化学演化 133
circulating variable 圆环动变量 76

cloud 云
  interstellar 星际云 23
  molecular 分子云 124
cluster of galaxies 星系团，见 galaxy cluster
CMB 见 cosmic microwave background
CO 见 carbon; monoxide
cocoon 茧 101
collisionless 无碰撞 99, 120
colour-magnitude diagram 颜色-星等图 43
component 组成部分 117
conserved quantity 守恒量 6
consistency 一致性 4
constants of motion 运动常数 120
convection 对流 29, 46
  in Sun 太阳中 33
coolants of gas 气体冷却剂 13；参见 dust grains; gas
core-collapse 核坍缩，见 supernova
corotation radius 共转半径 128
cosmic microwave background 宇宙微波背景 91, 136, 145, 147
cosmic rays 宇宙射线 45, 89, 93
cosmological simulation 宇宙学模拟 122
cosmology 宇宙论 145
crater 陨击坑 81
Cygnus A 天鹅射电源 A 62

D

dark energy 暗能量 145
dark globule 暗球状体 18

dark matter 暗物质 115—116, 119, 124, 145
  halo 晕 128
Davis, Ray 雷伊·戴维斯 45—46
decoupling, epoch of 退耦期 147
deflagration 暴燃 40；参见 supernova
deuterium 氘 26, 45
  to helium 变为氦 26
differential equation 微分方程 2, 82
Dirac, P.A.M. P.A.M. 狄拉克 88
disc 盘
  accretion 吸积盘 51—53
  stellar 恒星盘 119
displacement current 位移电流 3
Doppler-shift 多普勒频移 94—95
dust grains 尘粒 11, 23
  as coolant and regulator 作为冷却剂和调节器 13
  and interstellar chemistry 与星际化学 14
  and photo-heating of gas 与光子加热气体 13
dwarf 矮星，见 galaxy; star
dynamical friction 动力摩擦 129—130
dynamical model, equilibrium 平衡态的动力学模型 122

E

Earth 地球 77, 84
eccentricity 偏心率，见 orbit
Einstein, Albert 阿尔伯特·爱因斯坦 73, 91, 103, 112
Eddington, Arthur 亚瑟·爱丁顿 8, 112

Einstein's equations 爱因斯坦方程 104；参见 general relativity
electric and magnetic field 电场与磁场 100
electromagnetic 电磁(的)
    field 电磁场 99—100, 103
    waves 电磁波 113
electromagnetism 电磁 2
electron 电子 46, 88, 99
    acceleration 加速 35
    capture by nucleus 原子核捕获 36
    volt 伏 9
electrostatics 静电 102
elliptical 椭圆，见 galaxy; orbit
emission lines, width 发射线宽 69
energy 能量 5—6
    and mass 与质量 94
energy-momentum 能量-动量 103
entropy 熵 47
    maximum principle 最大熵原理 121
Eotvos, Baron Roland 厄缶男爵 103
evacuated annulus 真空环 78
event 事件 92；参见 special relativity
extra-solar systems 系外行星系 85

F

Fermat's principle 费马原理 107
Fermi 费米
    acceleration 加速 101
    process 过程 101
forward beamed 向前成束 95
fragmentation 碎裂 23

G

g-mode g 模式 47
Gaia satellite 盖亚探测器 112
galactic 星系(的)
    disc 盘 127
    fountain 喷泉 20
    wind 风 136
galaxy 星系
    anaemic spiral 弱臂 142
    cluster-dominant 星团主导 142, 144
    disc 盘 119
    elliptical 椭圆 119, 122, 140
    dark matter 暗物质 123
    dwarf 矮 125, 129
    spheroidal 椭球 116
    and gases, crystals 气体与晶体 121
    irregular 不规则 140
    lenticular 透镜状 118, 140, 142
    luminosity function 光度函数 116
    morphology 形态 115
    drivers of 驱动因素 139
    satellite 伴星系 141
galaxy cluster 星系团 138, 141
Galaxy, ours 银河系 115—116, 121
Galileo 伽利略 103
gamma-ray γ 射线 21, 68, 88
    burst γ 射线暴 89
    after-glow 余晖 89
    telescope 望远镜 88
gas 气体
    cold 冷气体 124, 142, 148
    compression 压缩 23
    coolants 冷却剂 13

索引

disc 气体盘 19, 117
  emission by 发射 15
  intergalactic 星系际 63, 136—139, 141—142, 148
  virial-temperature 位力温度 137—138, 141, 143, 148
Gastineau, Mickael 迈克尔·加斯蒂内奥 84
general relativity 广义相对论 59, 102—114
  and cosmology 与宇宙论 91, 145
  and deflection of light by Sun 与太阳引起的光线偏转 110
  and Mercury 与水星 73
  and stability of solar system 与太阳系的稳定性 84
  and strong lensing 与强透镜成像 109
  and weak lensing 与弱透镜成像 108
globular cluster 球状星团 42
  age 年龄 43
gold 金 38
gravitational 引力（的）
  field 引力场 103, 108, 113, 115, 123—127, 131
  effects of fluctuating 涨落效应 124
  lensing 透镜 70, 105
  strong 强 107
  weak 弱 108
  microlensing 微引力透镜 109
  redshift 红移 105
  waves 波 113—114
gravity 引力 102
  and star formation 与恒星的形成 23
  weak-field 弱场 104

H

half-life 半衰期 93
Hawley, John 约翰·霍利 61
heavy elements 重元素 123, 134, 136, 148
  abundance 丰度 135
  nuclei 核 23
Heisenberg uncertainty principle 海森堡不确定性原理 7, 36—37, 40
helium 氦 146
  burning 燃烧 26, 29
  flash 闪 29
  ignition 点火 29
  synthesis from hydrogen 用氢合成 23
Herbig Haro object 赫比格-哈罗天体 61
Hertz, Heinrich 海因里希·赫兹 3
hot spot 热斑 59
Hubble constant 哈勃常数 107
Hubble Space Telescope 哈勃太空望远镜 123, 139
Hulse, Russell 拉塞尔·赫尔斯 90
Hulse-Taylor pulsar 赫尔斯-泰勒脉冲星 90, 113；参见 pulsar
hydrogen 氢 15, 61, 146
  burning 燃烧 25, 23, 28—29
  cyanide 氰化 14
  molecule 分子 15

I

ice giant 冰巨星 80
inclination 倾角 73
infrared light 红外线 22, 54, 68, 127

intergalactic gas 星系际气体，见 gas
interplanetary space 行星际空间 112
interstellar 星际
   absorption and reddening 吸收与红化 11
   cloud 云 23, 26
   gas 气体 148
invariable plane 不变平面 73—74
inverse Compton process 逆康普顿过程 68
iodine 碘 38
ionization 电离 17
iron 铁 23, 26, 41, 134
isochrone 等龄线 43

J

jet 喷流 24, 61, 95, 98, 143
   driving 驱动 65
   high-efficiency 高效 66
   internal structure 内部结构 65
   one-sided 单侧 96
   relativistic 相对论性 64—65
   scalability 比例不变性 64
   variability 可变性 66
Jupiter 木星 77, 79, 81, 110

K

Kamiokande II detector 神冈探测器 46
Kepler satellite 开普勒探测器 86
Kuiper Belt 柯伊伯带 80—81
   Primordial 原柯伊伯带 80

L

Large Hadron Collider 大型强子对撞机 90
Larmor frequency 拉莫尔频率 102
Laskar, Jacques 雅克·拉斯卡 84
late heavy bombardment 后期重轰击 82
laws of physics 物理定律 1, 4
Le Verrier, Urbain 于尔班·勒威耶 73
lead 铅 38
lensing 引力透镜，见 general relativity
lenticular 透镜状，见 galaxy
LHB 见 late heavy bombardment
librating 天平动 76
light 光
   curve 光变曲线 66
   deflection by Sun 太阳偏转 110; 参见 infrared; ultraviolet
   wavelength 波长 3
linear equation 线性方程 104
lithium 锂 133
Local Group 本星系群 118, 129, 141
Lorentz, Hendrik 亨德里克·洛伦兹 91
Lorentz 洛伦兹
   covariance 协变性 91
   factor 因子 87, 101
   transformation 变换 92, 94
Lyman alpha 莱曼 α 137

M

M31 见 Andromeda nebula

M33 见 Triangulum nebula

Magellanic Clouds 大小麦哲伦星云 115

magnesium 镁 41, 134

magnetic field 磁场 21, 33, 58, 60, 65, 101
   of Earth 地球 35
   reconnection 重连 34
   tension 张力 60

magnetorotatonal instability 磁转动不稳定性 61

main-sequence 主序, 见 star

Mars 火星 77, 84

Maxwell equations 麦克斯韦方程组 91, 102—104

Maxwell, James Clerk 詹姆斯·克拉克·麦克斯韦 2

Maxwellian distribution 麦克斯韦分布 121

Mayor, Michel 米歇尔·马约尔 71, 85

Mercury 水星 73, 77, 81, 84

metal 金属 44

micro-quasar 微类星体 68—69, 88

microwave background 微波背景, 见 cosmic microwave background

Milky Way 银河系 141, 144

missing matter 消失的物质 137

mode 模式 46

molecular 分子
   clouds 云 125
     dissociation 离解 18
     hydrogen, formation of 氢分子的形成 14
     ring, giant 巨分子环 128, 144

momentum 动量 5—6

Moon 月球 2, 82, 131

moving clocks 运动的钟 93

MRI 见 magnetorotational instability

muon μ 子 46
   lifetimes 半衰期 93

N

N-body simulation N 体模拟 128, 131

Neptune 海王星 77

neutrino 中微子 6, 37, 38
   electron 电子 46
   mass 质量 46
   muon μ 子 46
   oscillation 振荡 46
   solar 太阳 45
   tau τ 子 46

neutron 中子 37
   capture by nucleus 原子核捕捉的 37
   star 中子星 39, 58, 90, 113

Newton, Isaac 艾萨克·牛顿 1, 7, 71, 82

Newtonian 牛顿(的)
   gravitational potential 引力势能 105
   physics 物理学 91

NGC 1275 143

nickel 镍 41

Noether, Emmy 埃米·诺特 6

Noether's theorem 诺特定理 6

non-linear equation 非线性方程 104

normal-abundance population 正常丰度恒星群 134

northern lights 北极光 35
nuclear 核(的)
　　burning 核燃烧 23, 25
　　disc 核盘 128
　　energy 核能 24
　　fusion 核聚变 24
　　reaction 核反应 25, 30, 40
nucleosynthesis 核合成 123

O

orbit 轨道 115
　　chaotic 混沌 82—85
　　eccentricity 偏心率 73
　　elliptical 椭圆 73, 76
Orion's cloak 猎户座斗篷 20
oxygen 氧 139, 143
　　burning 燃烧 41

P

p-mode p 模式 47
Paczynski, Bohdan 玻丹·帕琴斯基 89
parsec 秒差距 8
particle acceleration 粒子加速 34
Pauli exclusion principle 泡利不相容原理 7, 36—37, 40
Pauli, Wolfgang 沃尔夫冈·泡利 6
Perseus galaxy cluster 英仙座星系团 143
perturbation theory 摄动理论 73, 104
photo-dissociation 光致离解 36
photon 光子 102

Balmer 巴耳末 17
　　escape from star 恒星逃逸 31
　　infrared 红外线 22
　　ionizing 电离 69
　　trapped by dust 尘埃捕获 22
　　ultraviolet 紫外线 24
photosphere 光球 29, 31—32, 38
PKB 原柯伊伯带 80
planet 行星
　　birth 诞生 76
　　disturbed 受摄 73
　　extra-solar 系外 71
　　free-floating 自由游荡 110
　　properties 性质 72
planetary 行星(的)
　　core 核 77
　　nebula 行星状星云 30
　　systems, evolution of 行星系统的演化 71, 77
plasma 等离子体 20, 37, 95, 112, 128, 143—144
Pluto 冥王星 80—81
Poincare, Henri 亨利·庞加莱 82
Poisson fluctuations 泊松涨落 124, 126
positron 正电子 88
precession 进动 74—75
　　of SS433 disc SS433 吸积盘 62
pressure and gravity 压力与引力 146
Primordial Kuiper Belt 原柯伊伯带 80
principle of maximum entropy 最大熵原理 121
PSR B1913 + 16 90
pulsar 脉冲星 113
　　Crab 蟹状星云 39

and gravitational waves 与引力波 113

Q

QCD 量子色动力学 39
QSO 准星体 57；参见 quasar
Quantum Chromodynamics 量子色动力学 39
quantum mechanics 量子力学 6—7, 16, 64, 88, 94
quasar 类星体 55, 68, 91, 123, 137, 139, 147
 luminosity fluctuations 光度波动 107
quasi-stellar object 准星体 57；参见 quasar
quasiperiodic variability 准周期光变 69
Queloz, Didier 迪迪埃·奎洛兹 71, 85

R

radar 雷达 112
radiation, black-body 黑体辐射 28, 36
radio 射电
 galaxy 星系 62, 88
 loud 强 68—69
 quiet 弱 69
 telescopes 望远镜 102
 observations 观测 143
 waves 波 3, 63, 113
radioactive decay 放射性衰变 38
ram pressure 冲压 101
reconnection 重连 34；参见 magnetic field
red giant 红巨星，见 star
reddening 红化 13
redshift 红移 57, 146
refractive index 折射率 105
relativistic 相对论(性)
 jet 喷流 64
 particle 粒子 89
 most energetic 能量最高的 90
relativity 相对论，见 special relativity, general relativity
relaxation 弛豫 100
resonance 共振 75, 83, 85, 127
 mean-motion 平运动 76, 79
resonant 共振(的) 75
rest-mass energy 静止质量能量 64, 87, 93
rotation of our Galaxy 银河系自转 15

S

S0 117；参见 galaxy; lenticular
Sagittarius Dwarf 人马矮星系 131—132
Saturn 土星 77, 79, 81
SBS 0957 + 561 107
Schechter luminosity 谢克特光度 116
Schmidt, Maarten 马丁·施密特 57
Schwarzschild, Karl 卡尔·史瓦西 104, 112
Schwarzschild solution 史瓦西解 104
secular resonance 长期共振 76
seismic waves 地震波 47
seismology 地震学 48
semi-major axis 半长轴 73；参见 orbit

Shapiro delay 夏皮罗延迟 112
shear 剪切力 51
shepherding 引导 79
shock 激波 37, 90, 99, 127—128
    and particle acceleration 与粒子加速 98
shocked plasma 受震激等离子体 98
silicon 硅 23, 41
    burning 燃烧 26
silver 银 38
simultaneity 同时性 92；参见 event
SLAC 斯坦福直线对撞机 87
slow drift 缓慢漂移 124
smoke 烟尘 11；参见 dust grains
SN1987A 35, 38—39
SN Ia Ia 型超新星，见 supernova, deflagration
SNO detector 萨德伯里中微子天文台 46
solar 太阳(的)
    neutrinos 中微子 45
    system 太阳系 71, 82, 84, 90, 112
    young 年轻的 79
    wind 风 35
sound 声/音
    speed 速 121
    waves 波 127
space 空间 11
    of orbits 轨道 120；参见 action space
special relativity 狭义相对论 91—102；参见 event; rest-mass energy
    and beaming 与光子束 95
    and Doppler shift 与多普勒频移 94
    and events 与事件 92

and Lorentz transformation 与洛伦兹变换 92
and muon lifetimes 与 μ 子半衰期 93
and rest-mass energy 与静止质量能量 93
and superluminal motion 与超光速运动 96
spectra of stars 恒星的光谱 32
spheroid 球体 118；参见 bulge
spin of protons and electrons 质子和电子的自旋 15
spiral
    anaemic 弱臂 142
    arms 旋臂 117, 128
    galaxy 旋涡星系 124
    structure 旋涡结构 127, 136
    wave 螺旋波 77
SS433 61
standard candle 标准烛光 41
Stanford Linear Collider 斯坦福直线对撞机 87
star (恒)星
    ages 年龄 48, 134
    alpha-enhanced α 增丰 134
    binary 双星 48, 113, 126
    formation 双星的形成 24
    cluster 星团 126
    destruction 破坏 126
    formation 形成 24
    colour 颜色 28, 43
    corona 冕 32
    dwarf 矮星 29
    black 黑 26

brown 褐 26
white 白 7, 30, 39, 40, 53—54, 58, 105, 134
envelope 包层 25
evaporation from cluster 星团的蒸发 126
exploding 爆发 35
formation 形成 22, 141
    efficiency 效率 24
    rate 率 136
luminosity 光度 27
main-sequence 主序 26—27, 54
mass 质量
    key 关键 25
    minimum 最小 23
    massive 大质量 23, 147
merger 并合 49
neutron 中子 39
red giant 红巨 29
seismology 地震学 46
shell burning 壳层燃烧 28
stability 稳定性 25—26
surface 表面 30—32
temperature 温度 28
statistical physics 统计物理学 121
stellar 恒星（的）
disc 盘 117, 125
dynamics 动力学 119
halo see also 晕 128, 133; 参见 star
Sun 太阳/日 26
    core 核 33
    luminosity 光度 9
    mass 质量 8
    orbit 轨道 120
    seismology 日震学 48

super-massive black hole 特大质量黑洞 123, 143
superluminal motion 超光速运动 96; 参见 special relativity
supernova 超新星 9, 19, 30, 35, 123, 134, 140—142
    in AD 1054 1054 年 39
    bubble 超新星泡 20
    core-collapse 核坍缩 36
    deflagration 暴燃 40—41, 134, 146
    and electron acceleration 与电子加速 21
    and heavy elements 与重元素 38
    and mechanical energy 与机械能 38
    as standard candle 作为标准烛光 41
synchrotron radiation 同步辐射 101—102

T

tau particle τ 粒子 46
Taylor, Joe 约瑟夫·泰勒 90
temperature 温度 16
    virial 位力 137—138, 140—141
thermal 热
    conductivity 导热率 121
    equilibrium 热平衡 121
tidal field 潮汐场 131—132
tide 潮汐 131—133
time-domain astronomy 时域天文学 66—70
tornado 龙卷风 50
torque 扭矩 51, 74

transition region 过渡区 32
transport coefficients 输运系数 121
Triangulum Nebula 三角座星云 118
Trojan asteroids 特洛伊小行星 82

U

ultraviolet light 紫外线 17—18, 54, 56, 69
units 单位 8—9
uranium 铀 38
Uranus 天王星 77

V

vacuum 真空 11, 92
van Allen Belt 范艾伦辐射带 35
van der Hulst, Frank 弗兰克·范德赫斯特 15
Venus 金星 77, 84
virial temperature 位力温度，见 gas; temperature
viscosity 黏性 51, 61, 66, 121
  variability 变化 67

W

wavevector 波矢 94
white dwarf 白矮星，见 star
William Herschel telescope 威廉·赫歇尔望远镜 89
wind 风 30, 38, 49, 141
  solar 太阳风 35

X

X-ray X 射线 53, 56, 69, 138, 141, 143
  binary 双星 68
  from accretion disc 吸积盘发出 68
  sources 源 90
  telescope 望远镜 139, 141

Z

Zwicky, Fritz 弗里茨·兹威基 7
  Principle 原理 7—8

索引

James Binney
ASTROPHYSICS
A Very Short Introduction

Contents

    List of illustrations i

1   Big ideas 1

2   Gas between the stars 11

3   Stars 22

4   Accretion 50

5   Planetary systems 71

6   Relativistic astrophysics 87

7   Galaxies 115

8   The big picture 145

    Further reading 149

List of illustrations

1 A parsec **9**

2 A dark globule **12**
  © European Southern Observatory/Science Photo Library

3 Luminosity of stars as a function of surface temperature **27**

4 The planetary nebula Messier 57 **31**
  NASA and The Hubble Heritage Team (STScI/AURA)

5 The outer layers of the Sun **32**

6 Magnetic field lines **33**

7 Plasma draining down magnetic field lines **34**

8 Magnetic field lines brought together **34**

9 The globular cluster NGC 7006 **42**
  ESA/Hubble & NASA

10 Brightness versus colour for stars in a globular cluster **44**
  A. Sarajedini et al, 'The ACS Survey of Galactic Globular Clusters. I. Overview and Clusters without Previous Hubble Space Telescope Photometry', *The Astronomical Journal*, Volume 133, Issue 4, pp. 1658–1672 (2007), http://iopscience.iop.org/1538-3881/133/4/1658/, after W. E. Harris, *AJ*, 112, 1487 (1996). © AAS. Reproduced with permission

11 An accretion disc **52**

12 Temperature versus radius in an accretion disc **54**

13 The luminosity radiated by an accretion disc **55**

14 The temperature of an accretion disc around a black hole of mass 10^8 $M_\odot$ **56**

15 The luminosity radiated outside radius (r) for a disc around a 10^8 $M_\odot$ black hole **56**

16 Gas is lifted off a disc by DQ Her **59**
© Russell Kightley/Science Photo Library

17 Four snapshots of magnetic field lines in an accretion disc **60**

18 The Herbig-Haro object HH30 **62**
C. Burrows (STScI & ESA), the WFPC 2 Investigation Definition Team, and NASA

19 The geometry SS433 **63**

20 A radio image of the corkscrew jets of the binary star SS433 **63**
K. M. Blundell & M. G. Bowler, using the U.S. National Radio Astronomy Observatory's Very Large Array telescope; *Astrophysical Journal* 616, L159 (2004)

21 A radio image of the radio galaxy Cygnus A **64**
Image courtesy of National Radio Astronomy Observatory/AUI

22 An orbital ellipse of eccentricity $e = 0.5$ **72**

23 The inclination angle between the invariable plane and the plane of the planet's orbit **74**

24 The positions of two planets on similar orbits shown at a hundred equally spaced times **75**

25 A planet orbiting an (invisible) central star **78**
F. Masset (UNAM)

26 Angular distribution of emission from jets **96**

27 The geometry of superluminal expansion **97**

28 The apparent velocity of a blob **98**

29 Plasma coming fast from the left hitting slower-moving plasma on the right **99**

30 Computer simulation of a jet-inflated cocoon **101**
M. Krause, 'Very Light Jets II: Bipolar large scale simulations in King atmospheres', *Astronomy & Astrophysics*, 432, 45 (2005) reproduced with permission © ESO

31 A lens and a gravitational lens **106**

32 Weak lensing in the galaxy cluster Abell 2218 **108**
W. Couch (University of New South Wales), R. Ellis (Cambridge University), and NASA/ESA

33 The micro-lensing event OGLE-2006-BLG-109 **111**
Gaudi et al., 'Discovery of a Jupiter/Saturn Analog with Gravitational Microlensing', *Science*, 319, 927 (2008). © AAAS. Reproduced with permission via Copyright Clearance Center, Inc.

34 Galaxy luminosity function **117**
D.H. Jones et al., 'Near-infrared and optical luminosity functions from the 6dF Galaxy Survey', *MNRAS*, 360, 25 (2006)

35 An image of our Galaxy **118**
2MASS/J. Carpenter, T. H. Jarrett, & R. Hurt. Atlas image obtained as part of the Two Micron All Sky Survey (2MASS), a joint project of the University of Massachusetts and the Infrared Processing and Analysis Center/California Institute of Technology, funded by NASA and the National Science Foundation

36 Cartoon of disc, bulge, and halo **118**

37 Formation of a wake behind a massive body **130**

38 Tides without a moon **132**

39 A simulation of tidal tails forming **133**
J.L. Sanders & J. Binney, 'Stream–orbit misalignment–II. A new algorithm to constrain the Galactic potential', *MNRAS*, 433, 1826 (2013)

40 Stars formed in the first gigayear **135**
Recto-Blanco et al. (2014)

41 The Coma cluster **138**
(a) NOAO/AURA/NSF TBC; (b) S. L. Snowden USRA, NASA/GSFC

42 A dwarf galaxy and cold, star forming gas **142**
Geha et al., *Astrophysical Journal*, 653, 240 (2006), fig. 4

Chapter 1
Big ideas

In heaven as on Earth

Before Newton there was astronomy but not astrophysics. If legend is to be believed, astrophysics was born when Newton saw an apple drop in his Woolthorpe orchard and had the electrifying insight that the moon falls just like that apple. That is, a celestial body such as the moon does not glide on a divinely prescribed path through the heavens as Newton's predecessors supposed, but is subject to the same physical laws as the humble apple, which tomorrow will be a half-eaten windfall not worth picking up.

The power of this insight is that it allows us to apply physical laws developed in our laboratories to understand objects that exist on the far side of the universe. Thus Newton's insight enables us to travel in the mind across the inconceivable vastness of the universe to view a massive black hole at the centre of a distant galaxy from which radio telescopes have received faint signals.

Newton laid the foundations of astrophysics in another key regard: he showed that it is possible to obtain precise quantitative predictions from appropriately defined physical laws. Thus he did not only give a coherent physical explanation of observations that had already been made, but he *predicted* the results of observations that could be made in the future. To do this, he had to invent new mathematics—the infinitesimal calculus—and use its language to encapsulate physical laws. Since Newton's time most physical laws have taken the form of differential equations. A differential equation specifies a function by stating the rate at which it changes. The differential equation encapsulates what is universal about a given physical situation, and the initial conditions that are required to recover the function encapsulate what is particular to a specific event. For example the trajectory of a shell fired from a gun is the solution of Newton's equation $m \, d\mathbf{v}/dt = \mathbf{F}$, which is commonly abbreviated to $f = ma$ and relates the rate of change of velocity ($\mathbf{v}$) (the acceleration) to the force ($\mathbf{F}$) that is acting. Newton's equation applies to all shells and all falling apples, and to the moon. It is universal. The trajectories of the moon, the shell, and the apple differ by virtue of their initial conditions: the moon starts far from the Earth's centre and is moving exceedingly fast; the shell starts from the Earth's surface and is moving more slowly; and the apple also starts from near the Earth's surface but is initially stationary. These different initial conditions applied to one universal equation give rise to three completely different trajectories. In this way the mathematics that Newton invented became the means by which we identify what diverse events have in common and also what sets them apart.

It must hang together

James Clerk Maxwell, the only son of a prosperous Edinburgh attorney, early on displayed a great talent for mathematics and physics, and he made major contributions to the theory of gases and heat, and to the dynamics of Saturn's rings, but his greatest achievement was to extend the laws of electromagnetism by pure thought. He imagined a particular experimental set-up in which a alternating current flows in a circuit that includes a capacitor—a device consisting of two metal plates separated by a thin layer of insulator, which could in principle be a layer of vacuum. The current flows into one plate, charging it positively, and out of the other plate, charging it negatively. Maxwell applied to this circuit the rules for calculating the magnetic field generated by a circuit that had been developed by André-Marie Ampere. In 1865 he showed that these rules gave rise to completely different answers depending on how you applied them unless there was a current flowing between the plates of the capacitor, through the insulator. This result led Maxwell to hypothesise that a time-varying electric field generates a 'displacement current'. Mathematically, the hypothetical displacement current constituted an extra term in the differential equation that related a conventional current to the magnetic field that it generated.

The astonishing implication of the extra term in the equation was that it enabled the electric and magnetic fields to sustain each other without the participation of charges—until then an electric field was what surrounded a charged body and a magnetic field was what surrounded a current-carrying wire. But with the extra term a time-varying electric field generated a time-varying magnetic field, and Michael Faraday had already demonstrated that such a magnetic field generated a time-varying electric field. Thus the magnetic field regenerated the original electric field, without any charges being present! Could this amazing conclusion be correct, or was the extra term in the equation a foolish mistake?

Maxwell could calculate the speed at which the coupled oscillations of electric and magnetic fields would propagate through empty space, and that speed agreed to within the experimental errors with the measured speed of light. Maxwell concluded that his extra term was correct and that light *was* precisely mutually sustaining oscillations of the electric and magnetic fields. Because the wavelength of light was known to be extremely short (about 0.0005 mm) the frequency of the oscillations must be extremely high. Oscillations at lower frequencies would be associated with waves of longer wavelengths. In 1886 Heinrich Hertz generated and detected such 'radio' waves.

So Maxwell re-interpreted an old phenomenon, light, and predicted the existence of a completely new phenomenon by applying the conventional laws of physics to a thought experiment and arguing that the laws needed to be modified to ensure *consistency* of the theory. This was a ground-breaking step.

For ever and ever

We believe that the laws of physics have always been true: we have strong evidence that they were true a minute or so after the universe began 13.8 Gyr (gigayears) ago. They remained true as the universe evolved from exploding fireball through a cold, dark era to give birth to the first stars and galaxies, which are now being studied with huge telescopes. And they remain true to the present day.

Although the laws of physics have held steady over the last 13.8 Gyr, the universe has changed beyond recognition. Here again we have the Newtonian distinction between the laws of physics, embodied in differential equations, which are always and everywhere true, and the phenomena that they describe, which can change completely because the initial conditions for which we must solve the equations change radically.

Since the laws of physics are valid in every part of the universe, we can travel in our minds to distant galaxies. Because the laws of physics are valid at all times, we can travel in our minds back to the very beginning. The universal and eternal nature of the laws of physics enables us to become, in our imaginations, space-time travellers.

Astrophysics is the application of the laws of physics to everything that lies outside our planet. As such it is the child of other sciences but completely dwarfs them in its scope.

In the beginning was the Word

The universe is transitory, while the laws of physics are eternal. They were there before the universe started, and they structured the universe. The running of any particular experiment cannot be the same from day to day, because in the real world things change. Today it's colder than yesterday, and this fact *will* change how the experiment runs to some extent. The Earth's magnetic field is constantly changing direction, and this will affect the experiment to some extent. The Sun is growing older and increasing in luminosity, the moon is drifting way from planet Earth, and these facts will affect the experiment to some extent. In the real world nothing stays the same, but in the world of a physicist's mind there are laws that are eternally true, that never change. This fixedness isn't an accident and it isn't a mirage: it's an act of will. A physicist doesn't feel s/he understands a phenomenon properly until it has been traced back to a law that's eternally true.

If we pack all our equipment up and ship it to another country, to another latitude, the experiment will run differently, to some extent, because at the new location the Earth's magnetic field will be different, because the Earth's gravitational field will be different, because it will be hotter or colder, and the flux of cosmic rays through the laboratory will be different. But the laws of physics will be precisely the same. Again the sameness of the laws of physics here there and everywhere is an act of will: we will not rest until any difference between the way the experiment runs in the new location and in the old can be traced to some difference in the circumstances changing the solution we require to the immutable and universal laws of physics.

This insistence on explaining phenomena in terms of laws that are everywhere and always true doesn't only enable us to travel through space and time across the universe and back to the remotest times. It also equips us with three powerful weapons to take with us on our travels. These weapons are called energy, momentum, and angular momentum.

In 1915 Emmy Noether proved a crucial result. If the laws that govern a system's dynamics stay the same when the system is moved, or rotated, then as it moves or spins there is a quantity you can evaluate from its current position and velocity that will remain constant. We say the system has a 'conserved quantity'. The conserved quantity arising because the laws are the same everywhere is momentum, and the conserved quantity that arises because the system is indifferent to whether it is oriented east–west, north–south, or whatever other direction, is angular momentum. An extension of Noether's theorem is that if the dynamics is the same at all times then there is another conserved quantity, energy. Thus the universal and eternal nature of the laws of physics gives rise to three important conserved quantities, momentum, angular momentum, and energy. The constancy of these quantities is a big help when we are trying to understand a system that might be far away or back in the remote past.

In 1930 Wolfgang Pauli conjectured the existence of particles he called *neutrinos* that carried momentum and energy away during nuclear reactions. This conjecture was Pauli's reaction to experimental evidence that clearly showed non-conservation of energy and momentum. He conjectured the existence of unseen particles that ensure that energy and momentum *are* conserved. For a generation neutrinos were a pure speculation, but in 1956 they were finally detected. They are hard to detect because they have exceedingly small cross-sections $\sim 10^{-46}$ m^2 (metres squared), for interacting with anything. In the language of classical physics this means that a neutrino will collide with another particle only if it passes within $\sim \sqrt{10^{-46}}$ m = 10^{-23} m of the centre of that particle, a distance that is 100 million times smaller than the size of a proton. Actually quantum mechanics makes it meaningless to localize particles so precisely, so the real implication of the very small cross sections of neutrinos is that they have a very small probability of interacting at all. Nonetheless, neutrinos play a significant role in structuring the Universe.

More happens in heaven than on Earth

Our story started with Newton bringing the moon down to Earth by subjecting it to the ordinary dynamical laws. In the 1930s the eccentric Swiss astronomer Fritz Zwicky restored the primacy of the heavens to some extent by asserting that 'if it can happen it will'. That is, *anything* that it permitted by the laws of physics will happen somewhere in the Universe, and with the right instruments and a bit of luck we can *see* it happening. Zwicky's Principle indicates that it is profitable to think hard about what weird objects and exotic events are in principle possible. If your knowledge of physics is good, you will be able to calculate what the observable manifestations of these objects or events would be, and perhaps even estimate how often they occur. Then you can encourage observers to look for these events.

The classic example of this process is the identification of white dwarf stars. In 1930 Subrahmanyan Chandrasekhar was taking the long voyage from Bombay to Southampton to work at Cambridge University. He wondered how the then new and controversial quantum mechanics might have implications for stars. He showed that when a star such as the Sun runs out of fuel, it will cool and shrink to a tiny volume—the Sun will in its time shrink to the size of the Earth—and the pressure that maintains this fantastically dense object from collapsing under the intense force of its gravity is a pure manifestation of quantum mechanics: even though the star is cool, its electrons will be whizzing about at near the speed of light because if they didn't move so fast, the least energetic of them would be violating Heisenberg's uncertainty principle, which requires an electron whose location is rather certain to have very uncertain momentum. Moreover, the Pauli exclusion principle forbids two electrons from occupying the same quantum state, so most electrons are obliged to occupy quite energetic states because the states that just avoid conflict with Heisenberg's principle are all occupied.

When Chandrasekhar reached Cambridge excited about his wonderful theory, he was devastated to have it dismissed as nonsense by the dominant figure of British astrophysics, Sir Arthur Eddington. Eddington didn't accept Zwicky's principle, and he didn't accept that quantum mechanics, a seriously flakey theory developed to explain (after a fashion) the behaviour of atoms, applied to whole stars. But Chandrasekhar was right, and there are quite close to the Sun large numbers of these cold, hyperdense stars, sustained by a pure quantum-mechanical effect.

In Chapters 3 and 8 we will encounter other examples of successful predictions of amazing things made with the creative use of physics. Zwicky's principle works because the universe is so huge and varied that nature has conducted within it a stupendous number of experiments. Our planet is a very interesting place, but a restricted one, and if you want to understand the material world, you have sometimes to look up and away from it.

A note on units

Standard scientific units, kilograms, metres, seconds, etc., are matched to everyday human experience and when used in astrophysics require some very large numbers. For us a more convenient unit of mass is the mass of the Sun, $M_\odot = 2.00 \times 10^{30}$ kg, where 10^{30} is shorthand for 1 followed by thirty zeros.

When considering planetary systems a convenient unit of length is the *astronomical unit* (AU), the mean distance of the Earth from the Sun: $1\,\text{AU} = 1.50 \times 10^{11}$ m, in other words, 150 million kilometres. On galactic or cosmological scales even an AU is too puny to be handy, and the unit of distance is the *parsec* (pc), which

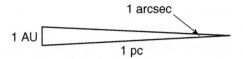

1. A parsec is the distance at which the Sun–Earth distance (1 AU) subtends an angle of 1 arcsec (1/3,600°).

is the distance at which a star that is stationary with respect to the Sun, when observed from the Earth, appears to move on the sky by one second of arc in a quarter year (Figure 1). From trigonometry 1 pc = 2.06×10^5 AU = 3.09×10^{16} m. The nearest stars lie about a parsec away, the centre of our Galaxy is 8.3×10^3 pc = 8.3 kpc (kiloparsec) away and on average one galaxy as luminous as ours is contained in a volume of $\sim$ 10 Mpc^3 (cubic megaparsec).

As our unit of time we usually take a year (1 yr = 3.16×10^7 s) although we usually have to deal with longer timescales: stars evolve over millions or billions of years. Hence we often write Myr for megayears or Gyr for gigayears, where
1 Gyr = 1,000 Myr = 10^9 yr.

A kilometre per second (km s) turns out to be a convenient unit of speed: the Earth orbits the Sun at $\sim$ 30 km s^{-1}, and the Sun orbits the Galactic centre at $\sim$ 240 km s^{-1}. Travelling at 1 km s^{-1} an object covers $\sim$ 1 pc in 1 Myr or 1 kpc in 1 Gyr. For example, in a gigayear the Sun covers $\sim$ 240 kpc while its path round the Galaxy has a length $2\pi \times 8.3$ kpc = 52 kpc, so it almost gets round five times in a gigayear.

The standard unit of power is the Watt (W) (roughly the rate of working when lifting a kilogram through 0.1 metres per second). A convenient astrophysical unit of power is the luminosity of the Sun $L_\odot$ = 3.85×10^{26} W. Utilities generally charge for energy by the kilowatt hour, or 3.0×10^{-28} $L_\odot$ yr. A supernova explosion (Chapter 3, 'Exploding Stars') injects into the surrounding interstellar gas $\sim 8.2 \times 10^9$ $L_\odot$ yr of energy.

While $L_\odot$ yr is a convenient unit of energy for astronomical objects, it doesn't suit atoms at all. When discussing atoms and subatomic objects the convenient unit of energy is an electron volt—(eV). 1 eV is the energy required to move an electron through a potential difference of 1 volt and is $10^{-53} \sim L_\odot$ yr. The photons our eyes can detect each carry $\sim$ 2 eV of energy, so in a year the Sun emits $\sim 10^{53}$ photons.

Chapter 2
Gas between the stars

The space between the stars is not completely empty although it is a vastly better vacuum than any that has been created on Earth: on average the gas near the Sun has 1 atom per cubic centimetre (cm^3) whereas air has $\sim 10^{19}$ atoms per cubic centimetre, so space near the Sun could be described as an ultra-high vacuum of 10^{-19} bar.

Interstellar absorption and reddening

This incredibly tenuous gas, mostly consisting of hydrogen and helium, manifests itself in many ways. One of the simplest and most important is through absorbing starlight. Actually, the starlight is not absorbed by the gas itself but by tiny particles of smoke that are embedded in the gas. Astronomers call these particles *dust grains*, but smoke is a much better name, for, as we shall see in Chapter 3, 'Life after the main sequence', they form in gases thrown off by certain stars precisely as soot forms in a burning candle or smoke forms in air drawn through a bonfire.

Naturally, the effectiveness with which dust absorbs starlight depends on the density of the dust, and therefore on the density of the gas within which it is embedded—it turns out that the mass of dust per unit mass of gas is roughly constant within our Galaxy. In a few directions the number of stars seen per unit area of the sky

2. A dark globule.

drops dramatically because in these directions there is a nearby dense cloud of gas, which obscures stars that lie behind it (Figure 2).

If you glimpse the Sun through the smoke of a bonfire, it appears redder than usual because blue light is more readily absorbed by small particles than red light. So red light from the Sun is more likely to pass through the smoke than the Sun's blue light. The physics of this selective absorption is that an antenna is an inefficient absorber of radiation with a wavelength much longer than itself: in the 1960s television aerials grew smaller when *ultra-high-frequency* broadcasting started (at ~ 0.3 GHz (gigaHertz)), and mobile phones shrank and ceased to have visible antennae when it became cheap to make electronics that could process radiation with wavelengths $\sim$ 15 cm. It turns out that the vast majority of interstellar grains are smaller than one micron (10^{-3} mm), so waves with wavelengths longer than a few microns are not much absorbed by dust. In fact we can see right into dense interstellar clouds by observing at wavelengths of few microns, longer than the $\sim$ 0.5 micron wavelength of visible light by a factor of about 4.

Since dust grains are efficient absorbers of blue and ultraviolet light, stars seen through interstellar clouds look redder than similar stars that have little gas in front of them. By comparing the colours of such pairs of stars we can determine the *reddening* of the redder star and thus the amount of dust and therefore gas along our line of sight to the star. It was in this way that the existence of interstellar gas was first established.

Dust the regulator

Dust grains play a crucial role in regulating the temperature, density and chemical composition of the gas. Electrons and protons that are whizzing about in interstellar space sometimes bump into a dust grain. The force of the impact sets the dust grain oscillating, and these oscillations cause the grain to radiate electromagnetic waves. In this way some of the the kinetic energy of the electrons and protons is converted into electromagnetic waves, which, as we shall see, are likely to escape from even a dense gas cloud. It follows from this that dust grains are major coolants of interstellar gas.

We have seen that dust grains absorb plenty of starlight, especially blue and ultraviolet starlight. Naturally the grains are warmed by this absorption, just as a sunbather is. And because their masses are extremely small, the absorption of a single photon can raise a grain's temperature dramatically. That is, a single photon can set a grain quivering quite violently. Any electrons or protons that adhered to the grain after colliding with it before the photon was absorbed are then violently shaken off, somewhat as water is shaken off a dog who has just been swimming. If the electrons and protons that are shaken off the grain move away from it faster than they were moving when they banged into the grain, overall the grain will have heated the interstellar gas. Thus grains can cool or heat interstellar gas depending on the intensity of starlight in the gas.

If the starlight is feeble, several protons and electrons can accumulate on a single grain between absorptions of photons. Then the protons may come near enough to each other as they jiggle around on the grain's surface to form a molecule of molecular hydrogen (H_2). Energy is released during the formation of an H_2 molecule and this is donated to the grain. When a photon next warms the grain the H_2 molecule may float free. So dust provides the main mechanism by which atomic hydrogen becomes molecular hydrogen.

Grains broker many other marriages too. Interstellar gas contains carbon, nitrogen, oxygen, and sulphur atoms in much lower abundance than hydrogen or helium atoms, but in significant abundances nonetheless. If a grain has both carbon and oxygen atoms sticking to it, a molecule of carbon monoxide (CO) is liable to form. If a grain carries a carbon atom and a nitrogen atom, a molecule of an even more poisonous gas, hydrogen cyanide (HCN), is liable to form because there's usually a hydrogen atom around to make up the party. In this way dust grains control the chemical composition of interstellar gas.

Emission by gas

Much of what we know about interstellar gas has been gleaned by detecting radiation from interstellar atoms and molecules. Hydrogen atoms consist of an electron in orbit around a proton and emit detectable radiation in two very different ways. One mechanism involves the radiation flipping the spin of the atom's electron relative to the spin of its proton: the energy of the atom is slightly higher when the spins are anti-aligned than when they are aligned, so an atom with anti-aligned spins can emit a photon by flipping the electron's spin. The wavelength of this photon (21 cm) is very, very much longer than the size of the atom—so the atom is an incredibly inefficient radiator of this long-wavelength radiation. In fact, left alone, an atom with anti-aligned spins is likely to stay that way for over a hundred million years before collapsing into the lower energy state. Fortunately, it can be very peaceful in the backwaters of interstellar space, and there are phenomenal numbers of hydrogen atoms out there considering making the transition, so if you tune a radio antenna to the magic frequency there's a strong signal coming from flipping atoms. The existence of this signal was predicted by Frank van der Hulst while he was a student in Nazi-occupied Leiden in the Netherlands. Detecting this signal was made possible by war-time work on radar, and in 1951 its detection was announced simultaneously by groups in the Netherlands, Australia, and the USA.

The detection of 21 cm radiation from atomic hydrogen dramatically improved our understanding of our Galaxy because for the first time we could get a clear picture of the rotation of our Galaxy. The radiation betrays the state of motion of the emitting atoms because the frequency at which an atom radiates is very precisely specified for an observer at rest with respect to the atom. If the atom is moving with respect to the observer, the frequency measured is shifted, to higher frequencies if the atom is approaching the observer, or to lower frequencies if it is receding (the Doppler effect).

Hydrogen molecules don't interact with photons that have less energy than ultraviolet photons, and such photons are scarce. So H_2 is very nearly invisible. This is a major problem for astronomers because roughly half the interstellar gas in our

Galaxy comprises H_2, and in many ways it is the most important half because it is the cold, dense half that is liable to turn into stars and planets. Fortunately CO provides a good tracer of H_2. A CO molecule is an 'electric dipole' because the oxygen (O) atom grabs more than its fair share of the electrons, making the carbon (C) end of the molecule slightly positive and the O end negative. Unless the gas temperature is extremely low, the CO molecules spin as they move, and a spinning electric dipole emits electromagnetic waves. These waves are emitted at very precise frequencies because quantum mechanics restricts the possible spin rates to discrete values: the molecule can have no spin (quantum number j = 0) or one unit of spin (j = 1), or two units, and so on. Moreover, a molecule can change its spin state by only one unit at a time, and when it passes from spin state j to $j-1$ it emits a photon that carries an amount of energy that is proportional to j. So all the frequencies of the photons emitted are multiples of the fundamental frequency associated with the transition $j = 1 \rightarrow j = 0$. These fundamental photons have wavelength 2.3 mm, and since wavelength is inversely proportional to frequency, the wavelength of the transition $j \rightarrow j-1$ is $2.3/j$ mm.

The probability that a given molecule is spinning at the rate j depends on the temperature of the gas—if the temperature is low, there is not much energy around and few molecules are either spinning fast or moving fast, whereas at high temperatures the molecules tend to both spin and move fast. Consequently the ratio of the number of molecules in the state j = 4, say, to those in the state j = 1 increases with rising temperature. It follows that as the temperature rises, so does the intensity of the spectral line with wavelength 2.3/4 mm relative to that with wavelength 2.3 mm. Hence measurements of several of these spectral lines enable us to determine the temperature of the gas.

In the 1970s it became possible to survey our Galaxy in the first few of these lines and thus to map the denser, colder part of the interstellar medium.

Nearby galaxies have long been mapped in both the 21 cm and 2.3 mm spectral lines. It is now possible to detect CO in very remote galaxies, and a global effort is underway to build a giant radio telescope, the Square Kilometre Array, which from 2020 will enable us to map the distribution of 21 cm emitting gas before the first stars and galaxies formed.

In its ground state atomic hydrogen does not absorb visible photons, but it does absorb energetic ultraviolet photons—those that carry more than 10.2 electron volts. Photons that carry 10.2 eV are called Lyman α photons and they play an important role in astronomy because they are very readily scattered by hydrogen atoms—the atom absorbs a photon and subsequently re-emits it in a different direction.

Photons that carry more than 13.6 eV of energy can strip the electron right out of a hydrogen atom. That is, they can ionize a hydrogen atom—convert it into a free electron and a proton. Subsequently, the proton is likely to capture a passing free electron, emitting a photon as it does so. The first photon emitted may carry only a small amount of energy because at first the electron may be only marginally bound. But once the electron has become trapped it is very likely to fall, like a drunkard who loses his/her balance on a staircase, deeper and deeper into the proton's electric field, emitting another photon as it slips down each step. Photons emitted as the electron falls to the next to lowest step on the staircase are known as *Balmer photons*. The least energetic of Balmer photons, Hα photons, have a beautiful pink colour and show up nicely in photographs of places where stars are forming because the hot stars in these regions are ionizing the gas around them, and thus preparing the ground for proton–electron recombinations.

Ultraviolet photons have a big impact on molecules as well as atoms because they can break molecules into their constituent atoms—*dissociate* molecules. In fact this is the main way in which molecules are destroyed—the molecules are formed on dust grains and destroyed by ultraviolet photons. Hence the chemical composition of interstellar gas hinges on the balance between the destructive power of ultraviolet photons and the catalytic action of dust grains. The higher the density of the gas, the more frequently atoms collide with dust grains and the larger is the proportion of atoms that are tied up in molecules. Moreover if the density of dust grains is high, the dust grains will absorb a significant fraction of the ultraviolet photons from hot stars before they can dissociate a molecule. So the fraction of the gas that is in molecular form increases rapidly with gas density.

If the density of gas in some region becomes high, a runaway situation can arise in which the density rises and rises almost without limit. This runaway occurs because as the density of the gas increases, fewer of the ultraviolet photons emitted by nearby stars penetrate deep into the cloud before being absorbed. But we have seen that grains are the major heat source of interstellar gas, and that molecules such as CO radiate energy. Hence a falling density of ultraviolet photons and a rising molecular fraction causes the gas to cool. Cooler gas exerts less pressure at the same density, so as it cools, the cloud contracts under the inward pull of its gravity. As the density rises, ultraviolet photons become still scarcer, the gas cools and contracts further. This runaway density leads to the formation of first the dark globules seen in Figure 2, and then stars.

The gas disc

Observations of the 21 cm line of atomic hydrogen and the 2.3 mm lines of CO reveal a thin layer of gas around the Galaxy's midplane. Beyond about 4 kpc (kiloparsec) from the centre, the gas is moving at close to the velocity of a circular orbit. The CO is more centrally concentrated than the atomic hydrogen, and more lumpy. It is also more concentrated towards the midplane, mostly being within

~ 40 pc (parsec) of the midplane rather than ~ 100 pc for the atomic hydrogen.

Every so often interstellar matter is blasted by a huge explosion. We will describe the objects (supernovae) that blow up in Chapter 3, 'Exploding Stars'. Here we consider the impact these explosions have on interstellar gas.

A supernova ejects between one and several solar masses at speeds of several thousand kilometres a second. The kinetic energy of the ejected gas is ~ 10^{44} J (Joules). For comparison, over its 4.6 Gyr life the Sun has radiated less than 6×10^{43} J, so we are talking about a serious quantity of energy.

The first gas to be ejected from the supernova slams into the essentially stationary ambient gas, compressing it, heating it, and jolting it into motion. Naturally, all this effort slows the ejected gas, so it is then hit from behind by gas that was ejected from the supernova a little later. This gas now slows, compresses, and heats. In this way a thick expanding shell of hot, compressed gas forms around the supernova. On both its inner and outer edges the shell is bordered by discontinuities in the gas velocity: on the outside it is slamming into stationary gas, and on the inside it is slowing gas streaming out of the supernova.

As the shell sweeps up more and more interstellar gas, it is cooled by expansion. If it expands undisturbed for long enough, its temperature will eventually drop to the point at which it is cooled by radiation faster than the two shocks are heating it. The denser the ambient gas is, the sooner this condition is reached because the luminosity of the shell is proportional to the product of its density and mass.

In Chapter 3, 'Star formation' we shall see that stars form in clusters, so the supernovae that mark the ends of the lives of massive stars cluster too. Hence a second supernova often goes off in the low-density region inside the expanding shell of an earlier supernova. Then the shell around the second supernova expands quickly through the low-density region and merges with the first supernova's expanding shell. Now we have a *supernova bubble*, which may well recruit further driving supernovae as it expands. In Orion there is a region of very active star formation with regular supernova explosions that is driving a wall of atomic hydrogen towards us at over 100 km s^{-1}—this wall is called *Orion's cloak*.

Some supernova-driven shells of fast-moving atomic hydrogen move away from the Galaxy's midplane and launch the gas onto an orbit in the Galaxy's gravitational field that carries the gas far from the plane. In fact roughly 10 per cent of the Galaxy's stock of atomic hydrogen is more than a kiloparsec away from the Galactic plane. Eventually this gas falls back to the plane, so one says that the supernovae are driving a *galactic fountain*.

When such a sheet of cool gas is shot into orbit, the way is clear for gas ejected by a supernova to flow clean out of the Galaxy. The gas is so hot that its electrons are rarely bound to ions, so it comprises free charged particles. In this condition we say a gas is a *plasma*. It is likely that this has been an important process over the life of the Galaxy and intergalactic space is rich in supernova ejecta.

If star-forming events are sufficiently common, adjacent bubbles will overlap. Even though supernova explosions are thought to occur in our Galaxy once in ∼ 50 yr (years) on average, supernova bubbles have overlapped to the extent that most of interstellar space is filled by them, with the denser interstellar gas squeezed into the narrow spaces between bubbles. The pressure (P), density (n), and temperature (T) of an ideal gas are connected by Boyle's law, $P = \text{constant} \times nT$, and the pressure exerted by the hot gas is roughly the same as that exerted by the cold gas, so the T and n of interstellar gas have an approximately constant product nT even though T varies from 20 Kelvin (K) to 2×10^6 K.

Supernovae accelerate electrons and ions to relativistic energies (Chapter 6, 'Shocks and particle acceleration'). These particles stream along the magnetic field lines that lace all interstellar space. From time to time a relativistic ion bumps into a nucleus of the interstellar gas, creating a gamma ray. The rate at which this happens in any volume is roughly proportional to the density n of the gas, so the intensity of gamma-ray emission is a valuable way to probe the density of interstellar gas.

We have seen that stars have a big impact on interstellar gas. Now let's take a look at how stars work.

Chapter 3
Stars

To this day our most important probes of the universe are telescopes that gather either visible photons or photons with only slightly longer wavelengths (infrared photons). At these wavelengths the night sky is entirely dominated by stars. We detect about a billion stars individually, as tiny unresolved points of light, and a billion billion more as contributors to the light coming from galaxies so distant that we cannot distinguish individual stars in the great agglomerations of stars that galaxies are.

So most of what we know about the Universe has been gleaned from a study of stars, and one of the major achievements of 20th-century science was to understand how stars work, and to understand their life-cycles from birth to death.

Star formation

Stars form when a cloud of interstellar gas suffers a runaway of its central density as discussed at the end of Chapter 2. After the density has increased by an enormous factor, of an order of a million million (10^{12}), photons emitted by atoms and molecules start to have trouble escaping from the cloud because they are scattered by molecules and dust grains after going only a small distance through the dense mass of gas and dust. When you pump up your bicycle tyres, the pump becomes warm from the work you do compressing air inside it. Similarly, as gravity compresses the gas of a collapsing cloud, work is done on the gas and the gas will warm if it cannot radiate the newly acquired energy. Once photons find it difficult to escape from the cloud, the work done by compression cannot be radiated in a timely manner and the temperature begins to rise. However, even as the temperature and pressure rise at the centre of the cloud, the crushing force of gravity increases too as more and more gas falls onto the core of the cloud. The consequence is a prolonged period of rising central temperature and density. If a cloud is sufficiently massive, the temperature and density are eventually sufficient to ignite *nuclear burning*, energy release by transmuting hydrogen to helium, and then helium nuclei into heavier nuclei such as carbon, silicon and iron. We discuss nuclear burning below.

When an interstellar cloud suffers a runaway increase in its density, it does not form one star but a whole group of stars. We do not understand completely this process of fragmentation, but it is an important empirical fact. Within the deforming interstellar cloud several regions of runaway density arise, each capable of seeding a star. The rates at which these seeds accumulate mass varies greatly, with the result that a few give rise to massive stars and many give rise to low-mass stars. The most massive stars have masses $\sim 80\,M_\odot$, and the masses of stars extend down to below the mass $\sim 0.01\,M_\odot$ at which a star is too faint to detect at any point in its life.

Since the original cloud was a heaving, swirling mass of gas, the seeds move with respect to one another. One aspect of this motion is that one seed will get in the way of gas falling onto another seed, so augmenting its own growth and suppressing that of its neighbour. Another aspect of the relative motion is that seeds often go into orbit one around another to form a binary star.

Elsewhere whole groups of seeds go into orbit around each other to form a gravitationally bound cluster of stars. In Chapter 7, 'Slow drift', we shall see, however, that small star clusters are not stable and tend to evolve into a binary and a series of single stars.

As the seeds accumulate mass and begin to resemble stars, their nuclear energy output becomes more and more significant for gas in the lower density parts of the original cloud. The more massive stars start to radiate ultraviolet photons, which heat low-density gas as we saw on page 13. In Chapter 4 we shall see that young stars are surrounded by bodies of orbiting gas called accretion discs, and that these discs eject jets of gas along their spin axes. These jets slam into and heat diffuse gas in the neighbourhood. The upshot of all this activity by young stars is that quite soon after the density in a cloud runs away at specific locations, most of the cloud's gas is heated up and driven away. Consequently, from a cloud containing, say, $10^4 \, M_\odot$ of gas, only $\sim 100 \, M_\odot$ of stars will form. This low *efficiency of star formation* enables galaxies like our own to go on forming stars at a fairly steady rate throughout the age of the Universe because it implies a low rate of conversion of interstellar gas into stars.

Nuclear fusion

Atoms consist of a tiny positively charged nucleus surrounded by one or more electrons that move on orbits that take them $\sim 10^{-10}$ m from the nucleus. Nearly all the atom's mass is contained in the nucleus, which is only $\sim 10^{-15}$ m across. When two atoms collide, the orbits of the electrons deform so the distribution of negative charge surrounding each nucleus changes, and the nuclei experience electrostatic forces that deflect them from their original straight-line trajectories. The upshot is that even a head-on collision of two atoms is unlikely to lead to a collision of the nuclei themselves because their velocities will be reversed by electrostatic forces before the nuclei have a chance to come into contact. In a sense an atom's electrons provide a sophisticated anti-shock packaging for the nucleus that carefully protects the nucleus in all but the most extreme collisions.

As the temperature rises at the centre of a forming star, the atoms whizz about faster and faster and the violence of their collisions steadily increases. Electrons are knocked clean out of atoms so more and more nuclei become bare. Still colliding nuclei are unlikely to come into physical contact because, as positively charged bodies, they repel one another electrostatically. But eventually the collisions are so violent that some colliding nuclei actually touch. At this point nuclear reactions start to take place.

The energy scale of nuclear reactions is a million times larger than that of the chemical reactions that power our bodies and our cars. So the release of energy by nuclear reactions in the core of a cloud is a game changer. The density is soon stabilized at the value at which nuclear reactions release energy at just the rate at which heat diffuses outwards through the now massive overlying envelope of gas; if the rate of energy release is slightly lower than the rate of outward leakage, the central pressure falls, the core collapses, the temperature and density rise, and so does the rate of nuclear reactions. Conversely, if nuclear reactions are releasing energy faster than it can diffuse outwards, the central pressure rises, the core expands, the temperature falls and so does the nuclear reaction rate. Thus nuclear energy release makes a star an inherently stable mechanism.

Key stellar masses

A star more massive than $0.08\,M_\odot$ now settles to the business of nuclear burning. Stars more massive than $0.08\,M_\odot$ but less massive than $\sim 0.5\,M_\odot$ burn hydrogen to helium but cannot ignite helium. Stars with initial masses in the range $0.5 - 8\,M_\odot$ burn hydrogen and then helium, but cannot ignite carbon. Stars initially more massive than $8\,M_\odot$ but less massive than $\sim 50\,M_\odot$ burn carbon to silicon and then silicon to iron. Iron nuclei are the most tightly bound so no energy can be obtained by transmuting iron into any other element—iron nuclei constitute nuclear ash.

Stars more massive than 50 $M_\odot$ become unstable and explode before they have reached the stage of silicon burning. We know they become unstable, but are not certain what the final outcomes of these instabilities are. We think most of the star's mass is ejected into interstellar space leaving only a black hole as marker of the star's existence.

Stars with initial mass smaller than 0.08 $M_\odot$ only get hot enough to burn deuterium to helium. Deuterium is an isotope of hydrogen in which the nucleus consists of a proton bound to a neutron rather than a lone proton. Deuterium like hydrogen was created in the Big Bang and is destroyed in stars. It is $\sim 10^{-5}$ times less abundant than ordinary hydrogen, so it does not take long for a star to exhaust this fuel. An object that is only burning deuterium is called a brown dwarf. When the deuterium is consumed the object will cool to become an almost undetectable black dwarf.

The main phase in the life of a star more massive than 0.08 $M_\odot$ is the burning of hydrogen in its core. Since three quarters of the original interstellar cloud comprised hydrogen, in this stage there is lots of fuel to burn, and, as a bonus, more energy per nucleon (a neutron or proton) is released when hydrogen is burnt than when any other nuclear fuel is burnt. The Sun has been burning hydrogen in its core for 4.6 Gyr and it is only half way through the process. For reasons that will become apparent when star clusters are discussed in Chapter 3, 'Testing the theory', we call a star that's burning the hydrogen in its core a *main-sequence* star (Figure 3).

The more massive a star is, the more quickly it depletes its stock of core hydrogen and the shorter its main-sequence lifetime. Massive stars are spendthrifts: the bigger the inheritance of fuel they have at birth, the sooner they are bankrupt by virtue of having consumed that fuel. Figure 3 quantifies this fact by showing the

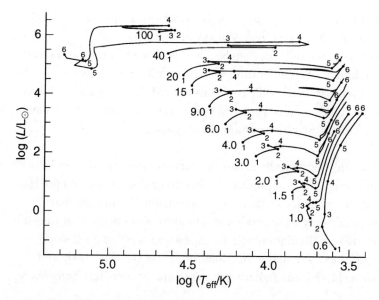

3. Luminosity plotted vertically in units of the luminosity of the sun as a function of surface temperature in degrees K for stars of various initial masses. The large number at the left end of each curves give the mass of a star in units of the solar mass $M_\odot$. While on the main sequence a star sits near the dot at the left end of its curve. It moves away from this dot when it has converted most of its core hydrogen to helium.

luminosities of stars of different masses as functions of surface temperature. During its main-sequence phase a star moves between the dot at the left end of its curve in this diagram and the point on the curve marked '2'. The big numeral gives the star's mass in solar masses. The vertical scale is logarithmic so there is a factor of a million in luminosity between the main-sequence points of stars with masses $0.6\,M_\odot$ and $20\,M_\odot$. Consequently, while a star of mass $0.6\,M_\odot$ will remain on the main sequence for 78 Gyr, nearly six times the age of the universe, a $20\,M_\odot$ star will be on the main sequence for just 8.5 Myr.

Figure 3 quantifies another key fact: the surface temperature of a main-sequence star increases with its mass. So massive stars are luminous, hot, and have short lives, while low-mass stars are faint, cool, and have long lives. When you heat a piece of metal strongly, it first glows dull red, then becomes yellow and white, and if you could raise its temperature even further it would glow blue. Hence hot stars are blue while cool stars are red. Figure 3 shows that all blue stars are massive, and we have seen that massive stars have short lives. So blue stars are always young.

The link between the colours and temperatures of stars reflects an important piece of physics. A *black body* absorbs any photon that hits it and emits a characteristic spectrum of radiation that depends only on the body's temperature and not on what the body is made of. Radiation with this spectrum is called *black-body radiation*. To a first approximation, a star is a black body and emits black-body radiation at the temperature of its photosphere.

Life after the main sequence

Once the hydrogen in a star's core has been consumed, hydrogen burning occurs in a spherical shell around the helium core and the core grows more massive, contracts and becomes hotter. In Figure 3 the star moves rather rapidly from point 2 to point 6 on its track, and we see that during this manoeuvre the luminosity increases, while the surface temperature falls. The rise in luminosity is most dramatic for low-mass stars, which have low main-sequence luminosities, and the decline in surface temperature is most pronounced for massive stars, which have high main-sequence temperatures. In fact, stars that have reached point 6 on their tracks all have rather similar temperatures, 2,000 K.

The reason the surface temperature drops as the luminosity rises, is that the increased flux of nuclear energy flowing from the hydrogen-burning shell puffs the star's enveloping gas into a bloated heaving body of gas in which energy is less transported by outward diffusion of photons that by convection. Convection is the process by which radiators heat rooms: hot air that has been in contact with the radiator rises, and is replaced by cooler air that falls down surfaces such as window glass and external walls that are unusually cool. The restructuring of the envelope into a bloated convective mass enlarges the star's photosphere, the sphere that emits most of the star's light. The swollen photosphere can radiate even the increased luminosity at a lower temperature than before. Since main-sequence stars are smaller than they will become once they have depleted their core hydrogen, they are called *dwarf stars*, and they will evolve into *red giant* stars.

At the point 6 in Figure 3 the helium in the star's core ignites. In stars more massive that $\sim 2\,M_\odot$ the ignition is quiescent, but in less massive stars the helium ignites explosively and we speak of the *helium flash*. The energy released in the flash causes the star's envelope to pulse violently and a significant portion of the envelope is ejected back to interstellar space. In stars more massive than $2\,M_\odot$ the regulatory mechanism described on page 25 functions properly and helium burning starts quietly. The onset of core helium burning causes the luminosity to drop slightly and the star to become slightly bluer.

The period of a star's life during which it quietly burns its core helium is the second most extensive period after that of the main sequence. Stars with initial masses up to $2.5\,M_\odot$ all spend $\sim 130\,\text{Myr}$ in this phase. At higher masses the duration of this period decreases rapidly with mass, and for a $20\,M_\odot$ star it is a mere $0.6\,\text{Myr}$.

Once the core helium has been burnt, helium burning shifts to a shell around the carbon core, beyond which hydrogen burning continues is an outer shell, and the star's luminosity rises rapidly. The star's envelope swells and instabilities frequently cause significant parts of it to be ejected into interstellar space. During this period these stars blow away most of their original mass in an increasingly powerful wind. As the gas flows outwards it cools and elements that form solids with high melting points condense into dust grains, so these stars have much in common with a Victorian factory chimney.

The rate of blow-off becomes a runaway process because as the star grows less massive, the power of gravity to hold gas in against radiation pressure in the envelope diminishes. Moreover, the luminosity of the star, and therefore radiation pressure grows as the quantity of gas that is blanketing the intensely hot core diminishes—the star's loft insulation is disappearing. Eventually the bottom of the envelope, where helium burning was taking place, lifts right off and the envelope becomes an expanding shell of gas around the star that is powerfully illuminated by the now naked core. The shell is ionized by photons from the core and glows brightly—the object is now a *planetary nebula* (Figure 4).

In the core nuclear reactions have ceased, so it gradually cools. It has become one of the white dwarf stars whose physics Chandrasekhar correctly outlined on the voyage from Bombay (page 7).

Stars initially more massive than $8\,M_\odot$ ignite carbon, burn it to silicon and then ignite that and burn it to iron. Since iron cannot be burnt, they are now obliged to replace heat that leaks out of their cores by contracting and thus releasing gravitational energy. Unfortunately, when a self-gravitating body contracts, its central temperature rises, and this rise in temperature soon proves fatal for the star—it suddenly blazes up into fireball and becomes a supernova.

The surfaces of stars

The density of gas in a star decreases continuously from the centre outwards, at first gradually but with increasing speed, although the density never falls precisely to zero. As the density falls, the distance a typical photon can travel before it's scattered or

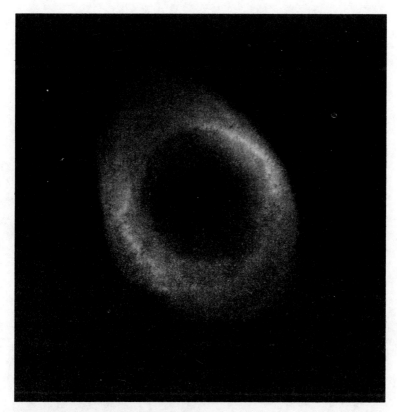

4. The planetary nebula Messier 57.

absorbed by an atom increases. At a certain radius this distance quite suddenly becomes comparable to the distance over which the density halves, and many photons can escape from that radius to infinity. The observed properties of the star are largely determined by the physical conditions in the spherical shell of this radius, the *photosphere* (Figure 5).

Photons of different frequencies escape from the star from radii that increase with the photon's propensity to be scattered by free electrons. Some photons have an unusually high propensity to scatter because they resonate with an oscillation of a common atom or molecule, and these remain trapped to the largest radii.

5. **The outer layers of the Sun. Sunlight comes from the photosphere. The temperature of the solar plasma reaches a minimum in the lower part of the chromosphere. In the transition region the temperature leaps from ~ 10,000 K to over a million degrees. The blisteringly hot corona extends very far out and is readily observed during a total solar ellipse.**

Hence the brightness of the star varies with wavelength and the star's spectrum contains spectral lines. The shape of these lines conveys information about radial gradients in density and temperature around the photosphere. Consequently, astronomers make huge efforts to obtain high-quality spectra for large numbers of stars. The precision with which the mass, radius, temperature and chemical composition of a star can be inferred from its spectrum is often limited by our ability to compute to the necessary accuracy the spectrum of light emitted by a star of particular mass, radius, etc.

Stellar coronae

The temperature of material in the Sun falls steadily all the way from the centre to the top of the photosphere—the visible surface—where it is about 4,500 K. Then, astonishingly, it starts to rise, at first slowly and then extremely rapidly—in the *transition region*, which is only 100 km thick, the temperature surges from ~ 10,000 K to over 1,000,000 K (Figure 5). Since heat always flows from hotter to cooler material, the corona must be heating the Sun, not the other way round. So what heats the corona? Outer space?

Convection carries much of the heat generated in the Sun's core on the last stage of its journey to the surface. Blobs of hot gas rise from 210,000 km below the photosphere, come to rest in the photosphere and there cool by radiating into space. Finally, they fall back to be reheated below the surface. Although convection is mainly an up-and-down process, in the photosphere gas does move horizontally before falling. So convection drives an unsteady circulation of gas.

The highly ionized gas in the Sun is a near perfect conductor of electricity because the many free electrons move easily in response to the tiniest electric field. Magnetic field lines freeze into and are swept along by such a conducting fluid, and the Sun's gas is magnetized. So the chaotic stirring of the Sun's surface by convection is constantly stretching and tangling the field lines that are embedded in the gas.

Magnetic field lines are analogous to elastic bands: there is a tension along the field line, and if a field line is stretched by the flow, the field grows stronger and its tension increases. In this case the fluid does work on the field; conversely, if the field line contracts, the field works on the fluid.

Adjacent field lines that are running in the same direction repel each other (Figure 6). If the field happens to be running parallel to the surface, this pressure pushes the field lines that are nearer the surface up and away from the field lines that run deeper down.

Gas cannot move across field lines, but it can flow down them, and once a field line has started to bow upwards, gas runs down the field line away from the crest of the bow. This flow diminishes the

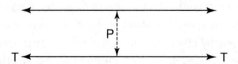

6. **Each magnetic field line is under tension and repels similarly directed field lines.**

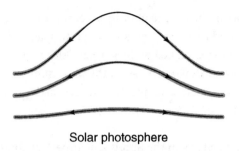

Solar photosphere

7. Plasma draining away from the crests of three upward bowing field lines.

8. When magnetic field lines that are moving in opposite directions are brought together, their oppositely directed sections cancel out, releasing magnetic energy, and the field settles to a different, 'reconnected' pattern.

weight that is bearing on bowed field lines, so they rise up some more, encouraging more gas to drain away from the peak, and soon a big loop of magnetic field is sticking out of the Sun's surface (Figure 7). Meanwhile, the dense gas in which the two ends of the loop is embedded flows over the Sun's surface in response to both convection and the systematic rotation of the Sun, and the field lines that make up the loop often become tangled in the sense that field lines that are moving in quite different directions are dragged close to one another. At this point the field *reconnects* somewhere in the corona as sketched in Figure 8.

When field lines reconnect, energy stored in the magnetic field is used to accelerate particles. Most of these energetic particles collide with nearby electrons and ions and lose their extra energy by heating nearby gas. So in the vicinity of a reconnection event the gas becomes extremely hot. Thus the searing heat of the corona is maintained by a constant flow of magnetic energy from the turbulent layer that is bounded by the photosphere through the 2,000 km thick chromosphere, a region of low-density gas that envelops the photosphere (Figure 5).

Much of the gas in the corona is too hot to be confined by the Sun's gravitational field, so it flows away from the Sun as the *solar wind*. About 60,000 km from the Earth the wind is deflect around us by the Earth's magnetic field. Electrons that have been accelerated to extreme energies in reconnection events above the photosphere, escape into the wind without losing much of their energy, and some of these particles become trapped in the Earth's magnetic field. These particles make up the *van Allen* radiation belts. They race at close to the speed of light from one magnetic pole to the other, exciting air molecules to glow as they approach the surface of the Earth near the north pole—this is the origin of the northern lights.

Exploding stars

Very occasionally a single star will become for a week or two as luminous as a whole galaxy of 100 billion stars. Such an event is called a *supernova*. In a galaxy like ours we expect a supernova to occur roughly every fifty years, although the last Galactic supernova to be observed was that found by Johannes Kepler in 1604—remnants have been found of supernovae that exploded in about 1680 and 1868, but the events themselves passed unnoticed. In February 1987 a supernova, *SN1987A*, was observed in the Large Magellanic Cloud, a small galaxy that is currently passing very close to our Galaxy and will eventually be eaten by it. This event provided by far the best opportunity to observe a supernova that mankind has so far enjoyed.

Supernovae are key cosmological tools because they can be observed out to vast distances. Consequently, major observational resources have in recent years been devoted to detecting and measuring large numbers of supernovae.

It turns out that two completely different mechanisms can generate a supernova.

Core-collapse supernovae

In the incredibly dense cores of stars that have burnt carbon to silicon, much of the pressure that resists gravity is provided by the electrons, which are obliged by the Heisenberg and Pauli principles (page 7) to whizz about much faster than they would do at the same temperature but a lower density. As a consequence they have so much kinetic energy that it can be energetically advantageous for them to become trapped inside a nucleus, lowering its charge and thus transforming it into the nucleus of the element before it in the periodic table. Each such capture reduces the number of electrons that contribute to the pressure opposing gravity.

As the core contracts, its temperature rises and the average energy of the photons in the ambient black-body radiation (page 28) rises. Eventually, this radiation contains a significant number of photons that are energetic enough to blast an atomic nucleus to pieces (*photo-dissociate* it). The gas of photons in the core makes a significant contribution to the pressure that resists gravity and each photo-dissociation reduces the pressure by withdrawing energy from the photon gas.

Hence the star is on a slippery slope: contraction drives up the temperature, which leads to more electrons being captured and more photo-dissociation of nuclei, which inevitably lead to further contraction. Within a few milliseconds the core is in free fall and cataclysm is inevitable.

As the central density rises, the atomic nuclei, so patiently assembled through the life of the star, are blasted apart. Most of the fragments end up as neutrons. Now the neutrons start to play the role that was earlier played by the electrons: they make a major contribution to the pressure by moving much faster than they would at the same temperature and a lower density because they have to conform to the Heisenberg and Pauli principles. Consequently, at some point the pressure within the core rises steeply with density and the core abruptly stops contracting, or 'bounces'.

Most of the star's mass lies outside this pressure-supported core and is falling inwards very fast. The inevitable result is a shock (Chapter 6, 'Shocks and particle acceleration') in which the inward falling material is brought to rest and violently heated.

The temperature and density are now so high that collisions within the plasma of electrons, neutrons and protons generate neutrinos in abundance. Because neutrinos have incredibly small cross sections for colliding with anything (page 6), they move significant distances between collisions even in the stupendously dense centre of the star. As a consequence the core become enormously luminous by radiating neutrinos rather than photons—it radiates photons too, but they are slow to diffuse outwards, so the neutrinos carry off energy much faster. So at this stage an enormous flux of neutrinos is flowing out through the envelope of the star, most of which is still falling onto the almost point-like core. A small fraction of the neutrinos collide with infalling nuclei, transferring energy and momentum to them. These transfers can be sufficient to reverse the inward motion of much of the envelope and blast it outwards in a great ball of fire.

Before the material of the envelope disperses into interstellar space, it is exposed to an intense flux of neutrons boiled off the neutron-rich core. Nearly all the neutrons are absorbed by atomic nuclei in the envelope, converting them into heavier nuclei. Usually the nucleus formed by the absorption of a neutron is highly radioactive and quickly decays to another nucleus, often by the emission of an electron and always with the emission of a photon. Hence radioactive decay becomes a significant source of heat within the dispersing envelope. Some nuclei absorb several neutrons one after the other, and undergo several radioactive decays. All elements that lie beyond iron in the periodic table were formed in this way—thus the nuclei of bromine, silver, gold, iodine, lead, and uranium were all created in supernova explosions.

As the fireball expands, its photosphere swells, so its optical luminosity rises. In the case of SN1987A the optical luminosity peaked three months after the core imploded—we know when the latter happened because the blast of neutrinos as the core bounced was detected. (To date SN1987A is the only supernova from which we have detected neutrinos.) Spectra taken at this stage show the material of the envelope to be fleeing the core at a few thousand kilometres per second. At this speed each solar mass of ejected material has 2×10^{43} J of kinetic energy, so the $\sim 5\,M_\odot$ of ejected material contains $\sim 10^{44}$ J of energy. This is just ~ 1 per cent of the gravitational energy released by the collapse of the core. This is a characteristic feature of supernovae: the spectacular explosion we detect and the profound impact that the event has on the interstellar medium are both powered by a tiny fraction of the energy that is actually released: 99 per cent of the energy is carried off by neutrinos that will never interact with anything, ever.

Eventually the expanding envelope becomes so diffuse that optical photons cannot be trapped for long within most of it. Hence its optical luminosity fades over a few weeks. As it expands and becomes more diffuse, dynamical interaction with the gas that was in the volume around the star becomes more important. In fact the gas density in this region is likely to be anomalously high, because before the star imploded as a supernova it was blowing off mass quite fast as a wind. The blast wave from the exploding star ploughs into the wind and shock heats it. Significant emission at radio wavelengths can arise at this stage.

Meanwhile, back in the core, much has been happening. We have seen that the collapsing core became very neutron-rich, and pressure generated by the neutrons caused the core to bounce and generate a burst of neutrinos that ejected much of the envelope. After the bounce, material continues to fall onto the stabilized core and a neutron star takes shape: this is a stupendously big atomic nucleus that's seriously neutron-rich. Within it neutrons dash around at mildly relativistic speeds, creating the pressure that resists gravity in just the same way that electrons resist gravity inside a white dwarf. The mass of this neutron star grows as material continues to fall onto it. As its mass grows, its radius shrinks (white dwarfs behave the same way), its gravitational field grows yet stronger, and the neutrons move ever faster to resist gravity. Remarkably, general relativity predicts that pressure is itself a source of gravity, so the harder the pressure resists gravity, the stronger gravity becomes. If the mass of the neutron star increases beyond a critical value, M_{crit}, gravity overwhelms the pressure generated by the neutrons and the object collapses into a black hole.

The precise value of M_{crit} is controversial because it depends on how matter behaves at nuclear densities. Although we believe we know what equations we need to solve, those of *Quantum Chromodynamics* (*QCD*), it is extremely hard to determine from first principles the required relationship between density and pressure. Moreover, we do not have good access to the relevant regime experimentally—the most massive nuclei on Earth contain only ~ 240 protons and neutrons and generate negligible gravitational fields. The experts are confident, however, that M_{crit} lies between $1.4\,M_\odot$ and $3\,M_\odot$, so some core-collapse supernovae leave neutron stars, while others produce black holes. The supernova recorded by the Chinese in 1054 left a neutron star (the *Crab pulsar*) that has been extensively studied. Our understanding of SN1987A implies that it involved the formation of a neutron star, but meticulous searches have failed to reveal a relic star of any kind.

Deflagration supernovae

We saw above that stars with initial masses smaller than 8 $M_\odot$ fail to ignite carbon and lose their envelopes, leaving a core of carbon and oxygen that gradually cools as a white dwarf star. If the star has no companion, that is the end of its story. The majority of stars do have a companion, however, and then the future can be much more exciting. If the companion was initially the less massive star, its evolution will occur on a longer timescale. Consequently, the companion will swell up and start blowing off mass at a significant rate after its companion has become a white dwarf. If the distance between the two stars is not too great, a significant part of the mass blown off by the companion will be captured by the white dwarf's gravitational field and form an accretion disc. Accretion discs of this type are studied by their X-ray emission (Chapter 4, 'Time-domain astronomy'). At the inner edge of the disc, gas transfers from the accretion disc to the white dwarf star, so the latter's mass increases.

As the mass increases, the star's radius decreases and its gravitational field becomes more intense. The Heisenberg and Pauli principles then decree that the most energetic particles speed up. Eventually some nuclei are moving fast enough to trigger the conversion of carbon into silicon. This conversion releases energy, which heats the star, so more nuclear reactions take place.

If the thermal motion of the nuclei were making a significant contribution to the pressure within the star, the star would respond to the heat input from nuclear reactions by expanding and cooling, both of which changes would slow the rate of nuclear reactions, and the system would be stable. But in a white dwarf the nuclei make an insignificant contribution to the pressure, which is dominated by the electrons. Consequently the density does not decrease as the nuclei are heated, and the rate of nuclear reactions spirals out of control. The technical term for the way the rate of nuclear reactions runs away is *deflagration,* a kind of slow explosion in which a front of enhanced temperature and reaction rate moves through the medium at a speed that is slower than the speed of sound.

Within a fraction of a second about a solar mass of carbon and oxygen has been burnt all the way to iron and nickel. The sudden release of energy by all this burning *does* produce enough pressure to overwhelm gravity and the star flies apart, leaving no gravitationally bound object where it was. A significant part of the dispersed material of the star consists of a highly radioactive isotope of nickel (^{56}Ni) which has a half life of 6.1 days. Gamma rays produced by the decay of ^{56}Ni to iron heat the dispersing material and cause it to glow brightly. It is by this glow that we detect deflagration supernovae. They are generally called *type Ia SNe* from the empirical classification of their optical spectra.

Type Ia SNe are important for astronomy in two different ways. First it proves possible to estimate the luminosity of a type Ia SN from the rate at which its brightness declines. Consequently, these objects can be used as *standard candles*: objects of known luminosity whose distances can be determined from their apparent brightnesses. Second type Ia SNe are major producers of iron since nearly all the original white dwarf is eventually converted to iron. Core-collapse supernovae, by contrast, produce a cocktail of heavy elements that is much richer in the *alpha elements*, which include carbon, silicon, magnesium, and calcium. Consequently, by measuring the abundance in a star of iron relative to the alpha elements, one can determine the relative importance of type Ia and core-collapse supernovae to the enrichment of the star's material. In Chapter 7, 'Chemical evolution', we shall see the value of this determination.

Testing the theory

The theory of stellar evolution requires as inputs a great deal of atomic and nuclear physics and involves both extensive numerical calculations and some assumptions about how turbulent fluids mix. Can we be sure that it is correct? We think it is fundamentally sound because it has now been possible to compare several aspects of what it predicts to the actual outcomes of observations.

Globular star clusters

Globular star clusters provided the classic tests of the theory in its formative years. Our Galaxy has about 150 star clusters that are very nearly spherical and quite compact—in a cluster such as NGC 7006 (Figure 9) several tens of thousand stars lie within $\sim$ 10 pc of the cluster centre—for comparison, within 10 pc of the Sun there are fewer than a hundred such stars. The feature of a globular

9. The globular cluster NGC 7006.

cluster that makes it an ideal test of the theory of stellar evolution is that, to an excellent approximation, its stars differ only in their masses: they have the same distance, age, and chemical composition. Moreover, the chemical composition of the stars can be estimated from the stars' spectra. For any conjectured age of the cluster and distance to it, the theory then predicts that the stars will lie on a curve called an *isochrone* in a *colour-magnitude* diagram, a plot such as Figure 10 in which the brightnesses of stars are plotted against their colour. For historical reasons, blue colours (implying hot surface temperatures) are plotted on the left of the diagram. The vertical brightness scale is always logarithmic, so changing the assumed distance to the cluster shifts the isochrone along which stars are predicted to lie up (for reduced distance) or down without distorting the isochrone's shape in any way. Changing the assumed age changes the isochrone's shape in computable ways. The age and distance are determined by finding the isochrone with the shape that best matches the observed distribution of stars and then finding the vertical position that produces the best match.

As Figure 10 illustrates, excellent matches between theory and observation can be produced in this way. Nevertheless, slight discrepancies do arise, and astronomers continue to refine the data and assumptions that go into stellar modelling to diminish these discrepancies. The ability of the theory to match data for many clusters leaves little doubt of its fundamental soundness, however.

A fascinating aspect of these fits is that our Galaxy's globular clusters prove to be extremely old—at one time the ages being derived were inconsistent with the age of the Universe. Since then refinements in our understanding of both the cosmic expansion and stellar evolution has yielded consistent ages. Clusters that have the lowest abundances of 'metals' (elements later in the periodic table than helium) tend to be the oldest, and even these have ages $\sim$ 12 Gyr that are not larger than the age on the Universe, 13.8 Gyr.

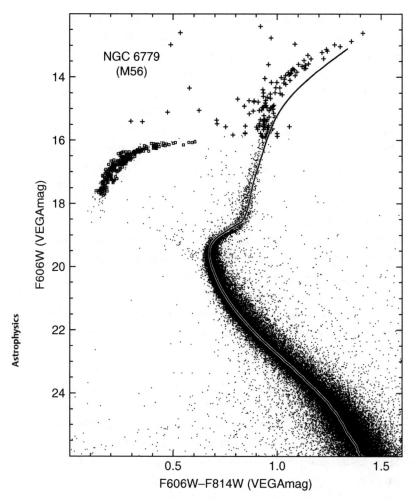

10. Brightness plotted against colour for stars in the globular cluster NGC 6779. Blue stars are on the left and bright stars at the top. The curves are theoretical isochrones (see text). The points from bottom right up to the sharp bend comprise the main sequence. Stars here are burning hydrogen in their cores.

Solar neutrinos

Every time two protons fuse in the Sun to make deuterium that shortly afterwards produces helium, a neutrino is produced that escapes from the Sun. Consequently, the Sun radiates neutrinos in addition to photons. In the 1960s Ray Davis set out to detect neutrinos from the Sun, arguing that their detection would be an important test of the theory of stellar evolution since they come to us direct from the Sun's energy-generating core and thus probe an entirely different region from the photosphere, from which we receive photons.

Davis's experiment involved processing tons of dry-cleaning fluid (tetrachloromethane CCl_4) down a mine—only by working in a mine could he exclude cosmic rays, which would generate a tiresome background signal. The CCl_4 was a convenient store of ^{37}Cl nuclei, which could become argon (Ar) after being hit by a neutrino. The argon had to be extracted and its quantity measured. After years of hard work, Davis measured only a third of the expected flux of neutrinos from the Sun. The scientific community was not very excited by Davis's failure: some wondered whether the experiment could detect Ar as efficiently as was claimed, others doubted the accuracy of the predicted neutrino flux.

These doubts were troubling because the experiment was sensitive to only a small minority of the neutrinos produced by the Sun: neutrinos produced by the fusion of protons don't have enough energy to transmute Cl into Ar; the more energetic neutrinos Davis hoped to detect came from other reactions whose rates are very temperature sensitive, and don't produce much of the Sun's energy. A small change in the rate at which heat is carried from the Sun's core could drastically lower the flux of these energetic neutrinos. So the experts re-examined their models of the Sun, but they were unable to reduce the flux of the higher energy neutrinos enough to be consistent with Davis's experiment.

Another issue was that Davis's experiment was only sensitive to one type of neutrino: there are three kinds of neutrinos, associated with electrons, muons, and tau particles. Davies's experiment could detect only electron neutrinos. This should not have been a problem as the nuclear reactions were expect to produce electron neutrinos. But could somehow two thirds of the emitted neutrinos pass through Davis's lab as undetectable muon or tau neutrinos?

From the mid-1980s two huge neutrino detectors were built that use either water (H_2O) or heavy water (D_2O) as the detector. A major advantage of these detectors is sensitivity to all three types of neutrino. One of these detectors, the *Kamiokande II detector* in Japan, observed the burst of neutrinos from SN1987A, and the other, the *SNO detector* in Canada, showed that when all three neutrino types are counted, the flux of neutrinos from the Sun is consistent with the original model predictions. This result from the SNO detector confirmed the idea that as a neutrino makes its way out of the Sun, it morphs from an electron neutrino into the other kinds of neutrino, with the result that roughly equal numbers of the neutrinos of each kind reach the Earth. Thus the astrophysics of the Sun had been correct from the outset, and the problem with Davis's experiment lay with particle physics.

The idea that neutrinos oscillate between the different types was first invoked to explain the outcome of solar-neutrino experiments, but later the process was studied in some detail using beams of neutrinos from nuclear reactors. It is a particularly important phenomenon because it implies that neutrinos have non-zero rest masses. Various experiments constrain the rest mass of the electron neutrino to a small value < 2 eV but neutrino oscillations establish that neutrinos have non-zero rest masses.

Stellar seismology
Stars like bells have natural frequencies at which they can oscillate. Each oscillation frequency is associated with a particular *mode* or type of oscillation. Bubbling associated with convection excites a star's modes of oscillation, and a great deal about the structure of a star can be learnt from measuring the spectrum of frequencies at which a star oscillates. Consequently, since about 1985 major observing programmes have monitored first the Sun

and later nearby, relatively bright, stars to determine their oscillation spectra.

A star's modes are of two basic types. The easiest to appreciate are the *p-modes*. These are analogous to the modes of an organ pipe: a standing sound wave is set up that involves alternating compression and rarefaction of the air in the organ pipe or gas in the star. P-modes are of less interest astrophysically than the other type of mode, *g-modes*. These are a little like ocean waves: when a less dense fluid (air) sits on top of a denser fluid (water), waves in which the surface of the denser fluid oscillates above and below its equilibrium level can propagate over the interface between the two fluids. Waves on the surface of the ocean are associated with a discontinuity in fluid density, but within the body of the ocean there is often a continuous gradient in density associated with salinity: salty water is denser than fresh water so in equilibrium more saline water underlies less saline water, and waves that distort the contours of equal salinity move across the ocean.

In a star, gas that has lower *entropy* underlies gas of higher entropy. Entropy is a measure of the thermal disorder in a fluid; it is increased by conducting heat into the fluid and decreased by extracting heat from it. It is distinct from temperature in that when air is compressed in a bicycle pump or the cylinder of a diesel engine, the air's temperature rises, but its entropy stays the same. Waves in which the surfaces of constant entropy oscillate up and down propagate around the star in the same way that waves in the surfaces of constant salinity propagate through the ocean. G-modes are standing waves of this type.

Oil companies prospect for oil by launching seismic waves with explosions and detecting the waves at remote sensors. A computer subsequently deduces the density and elastic properties of the rocks in the region from how the waves have travelled through the Earth from the source to the detectors. The pattern of frequencies of a star's oscillation modes is similarly sensitive to the density and rotation velocity of gas at different levels within the star, so with appropriate software one can constrain the density and rotation velocity within the star. These values can be compared with the predictions of theoretical models. The major uncertainty in the models is the star's age, which must in practice be estimated by matching models to the observational data, and the star's spectrum of normal modes most strongly constrains the age because as a star ages, its central concentration increases: the core contracts, growing hotter and denser, while the envelope expands. This evolution changes the pattern of oscillation frequencies.

The seismology of the Sun has demonstrated that models of the Sun that are founded on a wide range of data from nuclear and atomic physics work very well but not perfectly. Small discrepancies between the predictions of the models and the seismographic findings probably arise from limitations of the atomic data, or the stellar models. But it's just possible that they point to completely new physics being involved in the transport of energy out of stars: 'dark-matter' particles (Chapter 7) could become trapped inside stars and on account of their low propensity to scatter off other particles contribute disproportionately to the outward transport of heat.

Binary stars

At least half of all stars are members of a binary system, and the existence of binary stars is a major issue for the theory of stellar evolution because when the more massive, faster evolving, star swells up as it becomes a red giant star, its companion is liable to grab gas from the swelling envelope. This theft causes both stars to deviate from the evolutionary path we have described for single stars because mass is a key determinant of stellar evolution, and now each star's mass has become a function of time.

As material falls onto the more compact, less massive star, energy is radiated (Chapter 4). Some of this radiation heats the outer layers of the more massive star, and they may become hot enough to escape from the binary altogether as a wind.

Both the transfer of mass from one star to the other and loss of gas in a wind will change the binary's orbit and can draw the two stars closer together. If the stars do move closer, the rate of transfer or ejection of matter will accelerate, so these processes can run away and lead to the two stars merging. In fact the more massive star may envelop the less massive star in its swelling envelope even if the binary's orbit does not evolve up to that point.

Once the less massive star is inside the more massive star, its orbital motion will be opposed by friction. The envelope will be heated and the less massive star will spiral inwards. After a time the system will have become a single star, but neither the star's core nor its envelope could have been produced by the evolution of a single star.

In short, close binary stars constitute a Pandora's box of complexities, and attempts to understand the contents of this box are an active area of research.

Chapter 4
Accretion

When you empty the kitchen sink after washing up, the water usually swirls around the plug hole and leaves a column of air in the centre of the waste pipe as it runs away. This swirling action arises because the water has a tendency to conserve its angular momentum as it flows towards the waste pipe. Angular momentum per unit mass is given by the formula

$$L = rv_t,$$

where r is the distance from the point around which the fluid is flowing (the centre of the waste pipe) and v_t is the speed of the fluid's motion perpendicular to the direction to the centre. As the fluid flows inwards, r decreases, and v_t has to increase to keep L constant. Tornados (twisters) provide a more dramatic example of the same physics in action: as air is drawn towards the tornado's centre to replace warm, damp air that has floated up to high altitude, the air spins around the twister's centre faster and faster until it is moving so fast it can rip roofs off buildings, pick up cars, and generally wreak havoc.

The principle of angular momentum conservation was crucial for the creation of the disc of the Milky Way, within which the Sun is located, and in the formation of the solar system itself. In this chapter we will see that it plays a large role in many of the most exotic and luminous objects in the Universe.

Accretion discs

In all these systems gravity is sucking gas in towards some centre of attraction, which may be the centre of a galaxy, a star, or a black hole. That is, these objects are *accreting* gas, and conservation of angular momentum causes the accreting gas to spin around the accreting body as it moves in. If the gas is cold in the sense that its pressure is insufficient to provide effective resistance to the inward pull of gravity, the spinning gas forms an *accretion disc* in which the gas at each radius is effectively on a circular orbit around the centre of attraction.

Basic disc dynamics

In thinking about the structure of an accretion disc it is helpful to imagine that it comprises a large number of solid rings, each of which spins as if each of its particles were in orbit around the central mass (Figure 11). The speed of a circular orbit of radius r around a compact mass such as the Sun or a black hole is proportional to $1/\sqrt{r}$, so the speed increases inwards. It follows that there is *shear* within an accretion disc: each rotating ring slides past the ring just outside it, and, in the presence of any friction or *viscosity* within the fluid, each ring twists or *torques* the ring just outside it in the direction of rotation, trying to get it to rotate faster.

Torque is to angular momentum what force is to linear momentum: the quantity that sets its rate of change. Just as Newton's laws yield that force is equal to rate of change of momentum, the rate of change of a body's angular momentum is equal to the torque on the body. Hence the existence of the torque from smaller rings to bigger rings implies an outward transport of

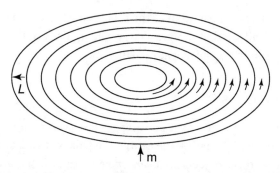

11. An accretion disc is imagined to comprise many solid annuli, all spinning at different rates about their common axis—the length of the curved arrows arrows is proportional to the local speed. The time it takes an annulus to complete a rotation increases outwards. Mass flows inwards and angular momentum L flows outwards through the disc.

angular momentum through the accretion disc. When the disc is in a steady state this outward transport of angular momentum by viscosity is balanced by an inward transport of angular momentum by gas as it spirals inwards through the disc, carrying its angular momentum with it.

As gas enters the outer radius of a ring, it has more energy than it has when it leaves the inner radius of the ring because its gravitational potential energy has decreased by twice the amount that its kinetic energy of rotation has increased. Hence each quantity of gas that passes through a ring deposits a quantity of energy in the ring. Moreover, the next ring in is doing work on our ring by torquing it in the direction of rotation at a rate that exceeds the rate at which our ring is working on the next ring out. Hence our ring also gains energy from the viscously driven flow of angular momentum through the disc. The energy gains from both the inward flow of matter and the outward flow of angular momentum heat the ring, causing its material to glow. This is why systems with accretion discs can be luminous astronomical objects.

To a good approximation each ring radiates as if it were a black body (page 28), so the spectrum of radiation from the accretion disc can be computed from the temperature $T(r)$ as a function of radius. The temperature of the ring of mean radius r is computed by balancing the rate at which energy is deposited against the rate at which radiation carries away energy. If the accreting body is gaining mass at the rate $(\dot{m})$, then the temperature in the disc is given by

$$T(r) = \left(\frac{GM\dot{m}}{2\pi r^3 \sigma} \right)^{1/4}. \qquad (4.1)$$

The temperature decreases outwards as the inverse three-quarter power of the radius and is proportional to the quarter power of the accretion rate.

Accretion onto stellar-mass objects

Figure 12 plots the temperature of the accretion disc around a solar-mass object that is accreting at a rate $\dot{m} = 10^{-8}$ $M_\odot$/yr, which are typical of the values inferred for many binary stars, where one star is dropping mass onto its companion (Chapter 3, 'Binary Stars').

The horizontal scale, which like the vertical scale is logarithmic, extends from the radius of a solar-mass black hole, marked $R_s \simeq 3$ km, up to radii beyond the orbit of Pluto (page 72). The temperature falls from 100 million degrees at R_s, a factor of several hotter than the core of the Sun, through the temperature $T_\odot$ of the Sun's surface at about the solar radius, down to about 100 K at the radius of the Earth's orbit. If the accreting object is a black hole, essentially the whole radial range plotted is physically relevant, while if the accreting object is a solar-mass white dwarf, only the part to the right of the line marked R_{wd}, where $T = 200,000$ K, is physically significant, and if the accreting object is a star like the Sun, the only relevant region is that to the right of

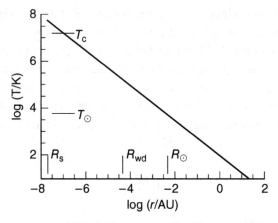

12. Temperature at radius (r) in an accretion disc around a compact object of one solar mass. An Astronomical Unit (AU) is the mean radius of the Earth's orbit (page 8). The accretion rate is assumed to be 10^{-8} $M_\odot$ yr^{-1}. Also marked are the temperatures T_c and $T_\odot$ at the centre and surface of the Sun and the radii R_s, R_{wd}, and $R_\odot$ of a solar-mass black hole, a typical solar-mass white dwarf star, and the surface of the Sun.

the line marked $R_\odot$, where $T = 4{,}700 \simeq T_\odot$. These temperatures are such that the disc will mostly radiate X-rays just outside R_s, soft X-rays, and ultraviolet light just outside R_{wd}, and mostly optical photons just outside $R_\odot$.

Figure 13 plots the luminosity radiated by the accretion disc from points outside the radius (r) on the horizontal axis. We see that the disc radiates only a few thousandths of a solar luminosity from beyond the Earth's orbit, about $L_\odot/2$ from beyond the radius of the Sun, about $60\,L_\odot$ from beyond the radius of a white dwarf and $\sim 100{,}000\,L_\odot$ down to the radius of a black hole. Taken with the conclusions we drew from Figure 12, it follows that when accreting at the rate 10^{-8} $M_\odot$ yr^{-1}, a solar-mass black hole will be an extremely luminous hard X-ray source, a white dwarf will be a luminous soft X-ray source, a main-sequence star will receive a significant boost to its luminosity from the accretion disc. In all these cases the portion of the disc that lies outside the Earth's orbit

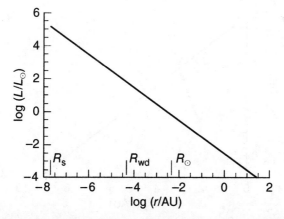

13. The luminosity radiated by an accretion disc around a solar-mass object outside radius r. Radii are given in units of the radius of the Earth's orbit and luminosities in solar luminosities. The accretion rate is assumed to be 10^{-8} $M_\odot$ yr^{-1}.

will contribute only infrared radiation and will be hard to detect against the luminosity from the inner portions of the disc and the accreting body. Nonetheless, this portion of the disc is of great importance as the site of planet formation (Chapter 5, 'Birth of planets').

Quasars

Figures 12 and 13 give results characteristic for stellar-mass black holes. A very different scaling is required for the black holes that sit at the centres of galaxies. These have masses in the range 10^6 to 10^{10} $M_\odot$ and when they are powering quasars they must have accretion rates of order 1 $M_\odot$ yr^{-1}. So these objects are $\sim 10^8$ times more massive that stellar-mass black holes and they accrete $\sim 10^8$ times faster. Since the characteristic radius R_s of a black hole is proportional to its mass, these holes are $\sim 10^8$ bigger in radius too. From equation (4.1) we see that at a given multiple of the black-hole radius R_s, the disc around a supermassive black hole has a temperature that is lower by a factor 100, and a luminosity which is a factor 10^8 greater than the corresponding values for a solar-mass black hole. Figures 14 and 15 reflect these facts.

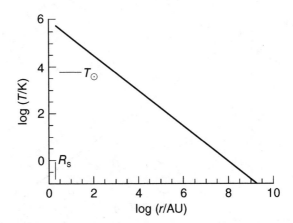

14. The temperature of an accretion disc around a black hole of mass 10^8 $M_\odot$ like those found at the centres of galaxies when the accretion rate is $1 M_\odot / \text{yr}$. R_s marks the radius of the black hole. The Sun's surface temperature $T_\odot$ is also marked for reference.

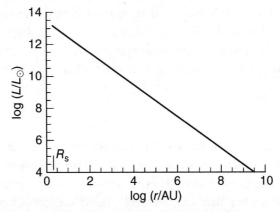

15. The luminosity radiated outside radius (r) for a disc around a 10^8 $M_\odot$ black hole that is accreting at a rate $1 M_\odot / \text{yr}$.

The characteristic radius of the black hole is now slightly bigger than the Earth's orbit and the temperatures reached there are only ~ 100,000 K so the bulk of the radiation will emerge in soft X-rays and ultraviolet light. The luminosity of such a system is staggering. A luminosity equivalent to that of our entire Galaxy of about 100 billion stars is radiated from the portion of the disc beyond 600 AU, so a factor ten larger than the radius of Pluto's orbit. A hundred times more luminosity emerges between that radius and the surface of the black hole.

On account of this prodigious luminosity, much of it emitted in the readily observed ultraviolet and optical bands, accreting supermassive black holes can be observed right across the Universe. When spectra of quasars were first obtained, astronomers failed to interpret them correctly because they were unprepared for the large shifts to the red of their spectral lines. This shift is quantified by the *redshift* z, which is defined by the relationship between the wavelengths at which light is emitted by the quasar and observed on Earth:

$$\lambda_{obs} = (1 + z)\lambda_{emit}.$$

Hence $z = 0$ implies no redshift, while $z = 1$ indicates that spectral lines are observed at twice the wavelengths at which they are emitted. In 1963 Maarten Schmidt dropped an intellectual bombshell by demonstrating that the spectrum of the object known as 3C-273 has a redshift $z = 0.158$. In the wake of Schmidt's paper, spectra of many other sources were shown to have even larger values of z, and the conventional interpretation of these redshifts was that they were caused by cosmic expansion. However, no galaxy was then known to have such a large redshift and many astronomers were sceptical that these sources could be as distant and luminous as the cosmological interpretation of the redshift required. Tens of thousands of the *quasi- stellar objects* (*QSO*) are now known, and some have redshifts in excess of $z = 7$!

Journey's end

After spiralling through the accretion disc almost to the radius of the accreting body, gas has to pass from the disc to the body, and the way it makes this transition impacts on the structure of the accretion disc behind it. Consideration of the case of an accreting white dwarf will reveal what's at issue. At the inner edge of the accretion disc gas is effectively on a circular orbit that skims over the surface of the star. The kinetic energy that keeps it in orbit is $E_c = \frac{1}{2}GM/r$ per unit mass, which is equal in magnitude to all the energy the gas would have lost if it had spiralled from infinity to r. Hence it is a prodigious amount of energy, and if the gas crashes into the star, this energy will be turned into heat in a flash. Since gas does ultimately crash into the star, a thin layer of exceedingly hot gas develops at the interface between the star and the disc. This *boundary layer* naturally boosts the emission of the system at the shortest wavelengths.

Recall that three quarters of the energy radiated by each annulus in the body of the disc arises from the difference between the rate at which work is done on that annulus by the annulus interior to it, and the rate it works on the next annulus out. The innermost annulus is in an anomalous situation because it does work on both the annulus outside it and the boundary layer, which spins slowly because it is braked by the star. So while the boundary layer glows unusually brightly, the innermost annulus of the disc glows more faintly than usual.

The situation we just described, in which the accretion disc extends right in to a slowly rotating solid star is not universal. Some accreting bodies (especially neutron stars and white dwarfs) have powerful magnetic fields embedded in them. Magnetic field lines emerge from the star, loop through the space around the star and re-enter the star at another location, just as field lines emerge from the Earth in the Arctic and plunge back into the Earth in the Antarctic. If a field-carrying star spins, the field lines whip round the star, and if the magnetic field is sufficiently strong they can exert significant forces on gas even at large distances from the star where they are rotating at a significant fraction of the speed required for a circular orbit. In these circumstances at some critical radius r_B gas moves from the accretion disc onto a passing field line (which is moving slowly relative to the orbiting gas) and

16. Gas is lifted off a disc by DQ Her.

then runs along the field line towards a pole of the star (Figure 16). In this situation there is no thin, hot boundary layer, but rather two hot spots, one near each pole of the star, where gas streaming along field lines suddenly crashes into the star.

As gas streams along field lines towards the star, is transmits its angular momentum through the field to the star, thus increasing the star's rate of spin. As the spin rate increases, the radius r_B decreases and the disc extends further in. The logical endpoint of this progression is when r_B shrinks to the radius of the star and the star is spinning so fast that matter on its equator is effectively in orbit. The star is then said to be spinning at *breakup*. In practice this extreme point is never reached, but accretion discs do spin neutron stars up to a significant fraction of the breakup rate.

Discs around black holes end in yet another way. Einstein's theory of general relativity decrees that the circular orbit around a non-rotating black hole that has the largest angular velocity has a radius of $6R_s$. Consequently, already at this radius we have an annulus that does work on the annuli either side of it, so is dimmer than would be the case in Newtonian mechanics. At $3R_s$ circular orbits become unstable and particles can simply spiral into the black hole without losing any energy: their energy is swallowed by the black hole. Hence there is no hot boundary layer around a black hole, and the innermost part of the accretion disc is fainter than our Newtonian calculations predict. The black hole is spun up in the same way that a white dwarf is.

Impact of the magnetic field

In Chapter 3, 'Stellar coronae', we saw that magnetic field lines are dragged along by conducting fluid, and are constantly exchanging energy with the fluid because a field line generates tension that pulls on the fluid. Because an accretion disc is in differential rotation, two initially neighbouring points soon diverge from one another unless they lie exactly the same distance from the disc's rotation axis (Figure 17). If these two points lie on the same field line, it follows that the field line is stretched, and the tension it generates increases. Now from Figure 17 we see that as the field line is stretched, the tension opposes the rotation of the mass element that's initially at a smaller radius, and pulls the outer mass element in the direction of rotation. The field will, in fact,

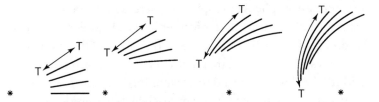

17. Four snapshots of magnetic field lines in an accretion disc being stretched by the disc's differential rotation. In each snapshot the star is shown at the bottom, and the earliest snapshot is on the extreme left. The disc rotates counter-clockwise. In that snapshot the field lines are short and run radially. By the final snapshot on the extreme right, the field lines are longer and are becoming tangential.

transfer angular momentum from the inner mass element to the outer mass element. But this is just what we expect viscosity to do! Moreover, by stretching the field line the differential rotation will strengthen the magnetic field, so no matter how weak the field was initially, it will amplify until it is strong enough to modify the dynamics sufficiently to limit the endless stretching of field lines. We have just described the essential physics of the *magnetorotatonal instability* (*MRI*), which in 1991 was proposed by Steve Balbus and John Hawley as the origin of viscosity in accretion discs.

Jets

The advent of the MRI not only solved the longstanding puzzle of why the viscosity within accretion discs was as large as the observations required, but more significantly it offered a clue as to the origin of the most remarkable aspect of accretion discs: jets. Wherever we have reason to believe a compact object is accreting, the observational data are dominated by *outflow* rather than inflow. A Herbig-Haro object such as that shown in Figure 18 is the classic example: the object's core is a forming star, but its most conspicuous feature is a pair of narrow jets along which very cold gas is racing at $\sim 200 \text{ km s}^{-1}$, way faster than the sound speed in the gas. On Earth aerospace engineers can only dream of replicating this trick.

Another remarkable example of jet formation is offered by the object SS433, which consists of a star rather more massive than the Sun in orbit around a black hole. The ordinary star is feeding gas to the black hole, and the gas spirals onto the black hole through an accretion disc. Two jets carry material along the spin axis of the accretion disc at over a quarter of the speed of light ($0.26c$, where c = 300,000 km s is the speed of light), yet the gas is cool enough to contain hydrogen atoms and emit the characteristic spectral lines of hydrogen. So again the gas in the jet is moving much faster than the sound speed.

18. The Herbig-Haro object HH30. Jets emerge either side from a disc accreting onto a proto-star. The disc itself is dark and flared; we see it as a silhouette against a bright background of scattered light.

In SS433 the spin axis of the accretion disc precesses around a line that lies within 11° of the plane of the sky (Figure 19), and at their bases the jets travel along the direction of the current spin axis. As each parcel of gas moves away from the black hole, it travels in a straight line but behind it the disc is precessing and launching fresh parcels along its new spin axis. So overall the jets spiral round a cone that has an opening angle of 40° (Figure 20).

Our final example of a jet is on an altogether grander scale: the radio galaxy Cygnus A shown in Figure 21. Two extremely thin jets

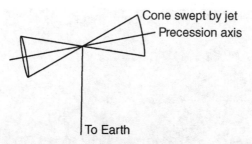

19. **The geometry of SS433. Oppositely directed jets sweep over the surface of a cone. This cone has an opening semi-angle of 20° and its axis is inclined at 80° to the line of sight to Earth.**

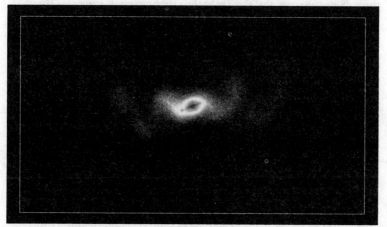

20. **A radio image of the corkscrew jets of the binary star SS433.**

of plasma emerge from the centre of a giant elliptical galaxy and 0.77 Mpc from the black hole slam into the intergalactic medium with incredible force, in the process accelerating electrons to stupendous energies (Chapter 6, 'Shocks and particle acceleration'). These energetic electrons emit the radio waves with which the system is imaged in Figure 21.

These three examples differ in many respects. The first involves accretion onto a young star while the last two involve accretion

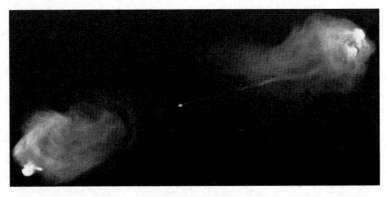

21. A radio image of the radio galaxy Cygnus A. Thin jets emerge from the galactic nucleus and slam into the circumgalactic gas more than 60 kpc away.

onto black holes. The first two are driven by stellar-mass objects while the last is driven by an object $\sim 10^8$ times more massive. The first involves completely non-relativistic jets, the second mildly relativistic jets and the third involves distinctly relativistic jets in the sense that particles in the jets have kinetic energies that exceed by a factor of a few their rest-mass energies (Chapter 6, 'Rest-mass energy'). Despite these great differences of scale each system consists of a pair of narrow jets which are carrying material away from the accreting object extremely supersonically. The ability of Nature to build systems that differ enormously in length and velocity scale but can nevertheless be roughly rescaled into each other implies that the physics of these systems is simple in the sense that it is restricted to electromagnetism and gravity, two theories that lack natural scales. By contrast, on Earth nearly all phenomena involve quantum mechanics, which introduces a scale through Planck's constant $h = 6.6 \times 10^{-34}$ J s. Stars, planets, the interstellar medium all have scales imposed on them through quantum mechanics and h. Jets seem to be a rare phenomenon that is independent of h. Since they are simple, you might imagine we understand how they form. Sadly, this is far from the case.

Driving jets

Although our understanding of jet formation is incomplete, we think we understand the basic principles. We have seen that we expect accretion discs to be laced by constantly amplified magnetic field lines. In Chapter 3, 'Stellar coronae', we saw that the Sun's surface is also laced by constantly stretched and twisted magnetic field lines, and that release of magnetic energy by reconnecting field lines in the low-density plasma above the photosphere drives a wind of plasma away from the Sun, past the Earth, and off into interstellar space. Something very similar must happen just above and below the mid-plane of an accretion disc, so the space above and below the disc is filled with gas that is too hot to be confined by the system's gravitational field and therefore flows away from the system. But in this case, unlike that of the Sun, the flow is somehow collimated into a narrow jet. It is likely that on account of the systematic rotation of the disc, the outflowing gas is confined by a helices of field lines that twist around the outflowing gas rather as a boa constrictor wraps itself around its prey to crush it to death. The tension in these field lines restricts the expansion of the ultra-hot gas in directions perpendicular to the disc's spin axis, so the gas instead expands in the direction of the spin axis, speeding up, and cooling down, as it does so. In this way a narrow column forms of gas that's moving much faster than its sound speed.

The process we have just described must happen at many radii simultaneously—this is a consequence of the scale-free nature of disc dynamics. At large radii the disc is cooler and rotating more slowly than at small radii, so we expect gas accelerated from these radii to achieve a lower ultimate flow velocity than gas accelerated from small radii. Hence we expect a jet to be a nested structure with a fast core surrounded by cylinders in which the flow velocity decreases steadily outwards. Even though the flow in the disc is never more than mildly relativistic (Chapter 6), ultra-relativistic jets are produced by accretion onto neutron stars as well as black holes. We do not understand how Nature achieves this feat.

High-efficiency jets

The emergence of nested jets from each side of an accretion disc requires a profound modification of the model of an accretion disc introduced in the section, 'Basic disc dynamics' at the start of the chapter. For that model was based on the assumption that the rate of flow of matter through the disc is the same at all radii. If gas leaves the disc at each radius to flow out in the jet, the rate of flow of matter through the disc must decrease as we move inwards. Moreover, previously, viscosity had to carry outwards angular momentum at the same rate that the flow of matter was carrying angular momentum in. When there is a jet, viscosity is not the only mechanism that takes away the angular momentum that is carried past radius r by the flow of matter: the jet carries angular momentum away from every point interior to r, so the viscous flow of angular momentum at r is smaller than it would be in the absence of a jet. Since this flow of angular momentum is responsible for three quarters of the heat input at r, the introduction of jets causes the disc to cool and to radiate less strongly. In fact, the energy output of the disc is being shifted from heat (in the form of radiation) to mechanical energy (in form of the jet's kinetic energy). Observations indicate that this shift can be almost complete, so almost all the energy released by the accretion disc emerging as kinetic energy.

Time-domain astronomy

We have studied accretion discs that have reached a steady state. However, the luminosities of real accretion discs tend to vary quite a lot, and a great deal can be learnt about a disc and the system in which it is located by monitoring the system's *light curve*, the luminosity as a function of time. More information still can be extracted if spectra taken at different times are available.

In some systems a significant quantity of gas is regularly dumped on a small region of an accretion disc. The matter is then quickly spread by differential rotation into an annulus of enhanced density. Then on a longer timescale viscosity carries angular momentum from the inner edge of the annulus to its outer edge, with the result that the inner edge moves inwards and the outer edge moves outwards. Hence the initially thin annulus of enhanced density becomes an ever broader region. Just as different radii within a steady-state disc heat to different temperatures, so the temperature of the inner edge of the annulus rises, and that of the outer edge falls, so the spectrum of radiation emitted by the whole annulus evolves away from a near black-body spectrum towards the kind of spectrum emitted by a steady-state disc. The timescale on which this evolution occurs is inversely proportional to the magnitude of the viscosity, which we did not need to know to derive the properties of a steady-state disc.

In some systems surges in luminosity are caused by a parcel of gas being deposited on the accretion disc as described above, but in others surges are caused by sudden changes in the magnitude of the viscosity. When the viscosity is low, matter takes a long time to move through the disc, so when the disc is in a steady state a given accretion rate onto the star corresponds to a large density of gas in the accretion disc. Conversely, when the viscosity is large the steady-state density in the disc is low but the luminosity is the same. Hence a sudden increase in viscosity within a disc that has reached a steady state causes matter to drain out of the disc onto the star faster than it is dropping onto the disc, and the luminosity surges before relaxing back to its original value as the high-viscosity steady state is approached.

What causes the viscosity to switch between high and low values? This is not well understood, but the mechanism is probably connected to the fact that the viscosity is generated by turbulence in the disc, and the turbulence is itself powered by viscosity. So in the low-viscosity state the turbulent eddies are small and the viscosity generated by these eddies is small, while in the high-viscosity state large eddies generate a large viscosity.

Many low-mass *X-ray binaries* regularly transition between a state in which the system has a soft X-ray spectrum and one in which it has a hard spectrum and a lower luminosity—the system, which contains an accreting black hole or neutron star, is alternating between low-hard and high-soft states. Often, but not always, relativistic jets squirt out as the system switches from its high-soft to its low-hard state. It is thought that this transition occurs when gas near the centre of the accretion disc is ejected along the spin axis, leaving a low-density region around the accreting body. In the high-soft state the dense central region radiates like a black body, while in the low-hard state the gas is too tenuous to create many photons itself. Instead it contributes to the system's luminosity by the *inverse Compton process* in which a relativistic electron collides with a photon. Just as a footballer's boot energizes the ball at a free kick, the electron can greatly increase the energy of the photon. In this way an infrared photon can become an X-ray or even gamma-ray photon and the spectrum of the object is hardened. X-ray binaries that sometimes fire jets are called *micro-quasars*.

The jets cause a micro-quasar's radio-frequency luminosity to increase by a factor of order 100—we'll discuss the physics of jets in Chapter 6, 'Rest-mass energy'. Consequently, at radio frequencies these sources are more variable than in X-rays, and when they are *radio loud* their radio emission is dominated by jets.

We have seen that the black holes that power quasars have radii R_s that are $\sim 10^8$ times larger than the radii of the stellar mass black holes that drive some X-ray binaries. The characteristic velocity of both types of systems is the same—a good fraction of the speed of light—so we expect quasars to vary on timescales that are longer by a factor $\sim 10^8$ than the timescales on which X-ray binaries vary. For example, a second in the life of an X-ray binary is equivalent to three years for a quasar, and the year or so between the changes of state on an X-ray binary is equivalent to 100 Myr for a quasar. Hence during our puny lifetimes we cannot expect to observe changes of state by quasars, but we do expect to find the population divided into radio quiet and radio loud populations. In fact before the connection between quasars and micro-quasars was recognized quasars had been divided into radio-quiet and radio-loud quasars, with more than 90 per cent of quasars being radio quiet. This proportion is in line with the fraction of their time during which micro-quasars are radio quiet.

At the blue end of the electromagnetic spectrum (X-rays or ultraviolet radiation depending on the nature of the accreting body) the brightness of many accreting systems flickers at a characteristic frequency—one speaks of *quasiperiodic variability*. Because quasiperiodic variability is concentrated at the blue end of the spectrum, which is dominated by emission from the inner edge of the disc, its characteristic frequency is thought to be the orbital frequency at the inner edge of the disc. Hence it tells us about the nature of the accreting body.

In micro-quasars quasiperiodic variability has a characteristic timescale of a millisecond, so in quasars the equivalent timescale is a day. The amplitude of these fastest fluctuations is small. On a timescale that's longer by a factor of a few hundred the X-ray luminosity of a micro-quasar can change by a good fraction of itself, and in quasars similar fluctuations happen on the timescale of a year. These fluctuations are of considerable diagnostic value.

For example, they can be used to estimate the mass of a quasar's black hole. When the luminosity of the accretion disc increases, gas in clouds that orbit the black hole at some distance is stimulated by ionizing photons from the inner disc to strengthen its optical and UV emission line emission. But there is a delay T between the luminosity of the accretion disc rising and the emission lines strengthening because the ionizing radiation takes time to cover the distance $r = cT$ from the quasar to the orbiting gas. The orbital speed v of the emitting gas can be estimated from the width of the emission lines, so using the formula $v^2 = GM/r$ for the speed of a circular orbit, we find that the mass of the black hole is cTv^2/G.

In conjunction with strong gravitational lensing (Chapter 6, 'Gravitational lensing'), fluctuations in the luminosities of quasar accretion discs can also be used to determine the scale of the Universe and to search for lumps of dark matter.

Chapter 5
Planetary systems

Astrophysics started with Newton's work on the dynamics of the solar system (Table 1), and work to understand how the solar system formed and evolved to its present state is at the forefront of astrophysics to this day.

In 1995 Michel Mayor and Didier Queloz announced the discovery of a planet orbiting the star 51 Pegasi, which is not unlike the Sun. Since then roughly a thousand extra-solar planetary systems have been discovered, and the process of understanding how these systems formed and evolved to their present states is having a profound impact on how we think about our own planetary system, and indeed our place in the Universe. Our understanding of the formation and evolution of planetary systems is developing rapidly, but we still don't know how unusual our system is.

Dynamics of planetary systems

Newton showed that if planets moved in the gravitational field of the Sun alone, their orbits would be ellipses (Figure 22). He was aware that this demonstration was only the start of a long journey towards understanding the complete dynamics of the solar system, because planets have non-zero masses and one needs to consider the impact of the mutual gravitational attraction of the planets. This undertaking was at the forefront of mathematical physics for

Table 1. The solar system The symbol $\mathcal{M}_\oplus$ denotes Earth's mass, while P_J is the period of Jupiter. We include Pluto although in 2006 the International Astronomical Union deprived it of the dignity of being called a planet on the grounds that it is merely a large Kuiper-Belt object.

Planet	$M/\mathcal{M}_\oplus$	a/AU	e	i	Period	$P:P_J$
Mercury	0.055	0.387	0.206	6.34	0.241	1:49.2
Venus	0.815	0.723	0.007	2.19	0.615	1:19.3
Earth	1	1	0.017	1.58	1	1:11.9
Mars	0.108	1.524	0.093	1.67	1.881	1:6.31
Jupiter	317.8	5.203	0.049	0.32	11.86	
Saturn	95.15	9.582	0.056	0.93	29.46	2.48:1
Uranus	14.54	19.19	0.047	1.02	84.02	7.08:1
Neptune	17.15	30.07	0.009	0.72	164.8	13.9:1
Pluto	0.002	39.26	0.245	17.1	247.7	20.9:1

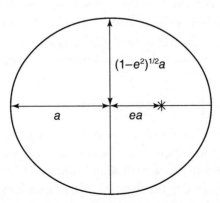

22. An orbital ellipse of eccentricity $e = 0.5$. The principal axes have lengths a and $\sqrt{1-e^2}\,a$. The centre of attraction, marked by a star, lies distance ea from the centre of the ellipse.

the following 250 years. It culminated in the work of Urbain Le Verrier (1811–77), who showed that Newtonian physics left a very slight anomaly in the orbit of Mercury. In 1916 an argument for the correctness of Einstein's brand new theory of general relativity was that it accounted for this anomaly very naturally.

In the last twenty years two developments have revived interest in planetary dynamics. The first was the availability of fast and relatively cheap computers, which made it possible for the first time to integrate the full equations of motion for billions of years, and the second was the discovery of extra-solar planetary systems, which are often dramatically different from ours and got astrophysicists wondering why that is so, and what it has to tell us about the solar system.

Disturbed planets

Since the masses of the planets are much smaller than that of the Sun, the natural approach to planetary dynamics is *perturbation theory*: we put the planets on the orbits they would have if they were all massless, and ask how a given planet's orbit will evolve in response to the force on it from the other planets. In this scheme the orbit of a planet is at all times one of Newton's elliptical orbits around the Sun, but the orbit in question slowly changes in response to perturbations from other planets.

The key numbers quantifying an orbital ellipse are its *semi-major axis a*, which controls the orbit's energy ($E = -GM/2a$), its *eccentricity e*, which describes the shape of the ellipse (Figure 22), and its *inclination i*, which is the angle between the plane of the ellipse and the *invariable plane*, an imaginary plane that is defined by the solar system's angular momentum (Figure 23). Table 1 lists these quantities for the Sun's planets. We use perturbation theory to compute how these numbers change over time.

Angular momentum is a crucial quantity in planetary dynamics, as in the dynamics of accretion discs (Chapter 4, 'Basic disc

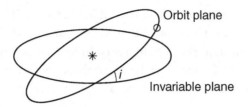

23. The inclination i is the angle between the invariable plane and the plane of the planet's orbit.

dynamics'). For a fixed semi-major axis a, the angular momentum

$$L = \sqrt{1-e^2}\sqrt{GMa} \qquad (5.1)$$

is largest when the eccentricity $e = 0$ and the orbit is circular, and decreases to 0 as e tends to unity.

We imagine that the mass of each planet is spread out along its orbit, so each orbit becomes an elliptical wire of slightly non-uniform density—the wire is densest where it is furthest from the Sun because at this point the planet moves most slowly (Figure 24). These wires attract each other gravitationally, and by virtue of the fact that they are elliptical and do not lie in the same plane, they exert torques (page 51) on one another as indicated in Figure 24. Torque gives the rate of change of angular momentum (Chapter 4, 'Basic disc dynamics'), so planets exchange angular momentum. To the extent that it is a valid approximation to replace planets by elliptical wires, planets do not exchange energy, so each planet's semi-major axis a is fixed while its eccentricity e changes.

If the planets had negligible mass, the orientation of the long axis of each planet's ellipse would remain fixed in space. If the effect of the mass of the planetary system were the same as that of a thin axisymmetric disc of matter in the invariable plane, the long axes of the planetary ellipses would rotate slowly in the sense opposite to that in which in which planets rotate around their ellipses. This backwards motion of the long axes is called *precession*.

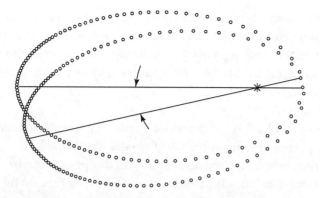

24. The positions of two planets on similar orbits are shown at a hundred equally spaced times. We represent each planet by an elliptical wire with mass density proportional to density of points drawn here. The arrows show the direction of the torque experienced by each ellipse due to the gravitational pull of the other ellipse.

In this axisymmetric model of the planetary system, each planet has its own precession frequency. Consequently, the torque that one planet applies to another keeps changing sign because over time the angle between the long axes of two planet's ellipses is as often such that planet 1 transfers angular momentum to planet 2 as it is such that the flow of angular momentum is from 2 to 1. In these circumstances the eccentricity of each planet's orbit oscillates slightly but nothing more interesting occurs. In particular, the angle between the long axes of two planets' ellipses constantly increases.

It can happen that the precession frequencies Ω_1 and Ω_2 of two planets are nearly *resonant* in the sense that

$$n_1 \Omega_1 \simeq n_2 \Omega_2, \tag{5.2}$$

where n_1 and n_2 are small integers. Then planet 1 can transfer angular momentum to planet 2 for a time long enough that the eccentricities of both planets change significantly. As a planet's eccentricity changes, so does the its precession rate, so the rate at which the angle between the long axes of the ellipses increases becomes non-uniform. In fact this angle can oscillate instead of continually increasing. We say that the angle between the ellipses is *librating* rather than *circulating* as it does when no resonant condition (5.2) holds.

Resonances like that just described are the key to understanding many phenomena in any branch of physics where small disturbances are involved because a small disturbance can be important only if it acts in the same way for a long time. In the absence of a resonance, the sense of the disturbance is constantly changing, so its time-averaged effect is zero; a resonance gives a weak disturbance the opportunity to act in the same sense for a long period, and thus to effect significant change.

A resonance that arises when we replace the planets by elliptical wires is called a *secular resonance* to distinguish it from a more fundamental resonance between the frequencies Ω_ϕ at which the planets move around their ellipses—in the wire model this motion has been averaged away. Two planets are in a *mean-motion resonance* when there are small integers n_1 and n_2 such that

$$n_1 \Omega_{\phi 1} = n_2 \Omega_{\phi 2}. \qquad (5.3)$$

Planets that are in a mean-motion resonance can exchange energy as well as angular momentum. Hence their semi-major axes a can change in addition to their eccentricities.

Birth of planets

A very young star is always surrounded by a disc from which it accretes matter. As the mass of the star increases, the temperature rises in its core (Chapter 3, 'Star formation'), and if the star's mass exceeds $0.08\,M_\odot$, nuclear burning starts up there (Chapter 3, 'Nuclear fusion'). As the luminosity of the star grows, its radiation heats the surrounding disc, and the warmed gas tends to escape from the gravitational field of the young star back into interstellar space. Particles of dust in the disc are too massive to escape, so the ratio of dust to gas in the disc increases as the young star warms its disc. In the increasingly dusty disc, dust particles collide and merge to form bigger particles. Eventually the most massive dust particles are kilometres in diameter and their gravitational fields are strong enough to deflect significantly the velocities of nearby gas and dust. An *asteroid* has formed.

Gradually asteroids collide and merge to form bigger and bigger asteroids. The self-gravity of the most massive asteroids pulls them into nice spherical balls, which become radially structured as dense material sinks and less dense material rises. These balls are planetary cores.

If a very massive core forms sufficiently early on, before all the gas has been dissipated at its radius, the core may trap some of the gas in its gravitational field. This is how the massive outer planets Jupiter, Saturn, Uranus, and Neptune formed, while the rocky inner planets Mercury, Venus, Earth, and Mars failed to acquire significant quantities of gas because their gravitational fields are not strong enough to retain hydrogen and helium at the relatively warm temperatures prevailing in the inner solar system.

Evolution of planetary systems

Early on, when no dust particle has a dynamically significant gravitational field, everything, gas and dust, is on nearly circular orbits in the system's invariable plane. Later the gravitational field of each young planet drives a spiral wave through the disc (Figure 25). The planet is gravitationally attracted to the nearby parts of the spiral overdensity. The inner region pulls it in the direction of rotation so it acquires angular momentum from this region, while it is pulled back by and gives angular momentum to the outer region. A detailed calculation shows that it loses more angular momentum than it gains. If the planet is quite massive,

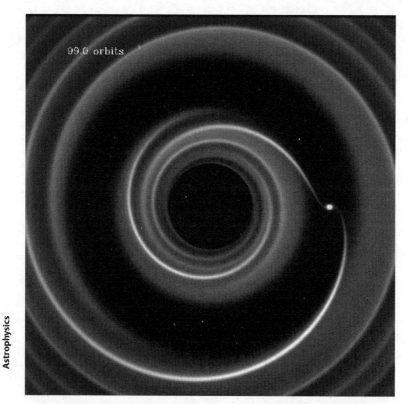

25. A planet (right centre) orbiting an (invisible) central star excites a spiral wave in the surrounding gas disc. Through this wave the planet gets angular momentum from the disc inside its orbit, and looses angular momentum to the exterior portion of the disc. These angular-momentum transfers create a region of near zero density around the planet.

these angular-momentum exchanges cause an annulus of low density to form around the planet. Material that was in this *evacuated annulus* has been either swept into the planet or pushed to beyond the edge of the annulus.

When an orbiting body loses angular momentum, it moves inwards. So the planet moves slowly towards the star. The material beyond the inner edge of the evacuated annulus would move in too as a result of surrendering angular momentum to the asteroid, but so long as plenty of gas is present, viscosity transfers angular momentum out through the disc fast enough to replace the angular momentum it has lost to the planet. Similarly, beyond the outer edge of the evacuated annulus, viscosity carries away the angular momentum provided by the asteroid and prevents the outer edge moving out. So the planet and its evacuated annulus moves slowly inwards *shepherding* the material of the disc so it remains in good gravitational contact with the planet at a safe distance on both sides of the evacuated annulus.

The rates at which planets drift inwards are not all the same, so the inner edge of the evacuated annulus of one planet can come into contact with the outer edge of the evacuated annulus of another planet. Numerical simulations of gas discs with embedded planets indicate that two planets are then likely to fall into a mean-motion resonance (see equation (5.3)) and become able to exchange both energy and angular momentum. It turns out that these exchanges lock the planets into the mean-motion resonance. Thus the inner planet picks up energy and angular momentum from the dust and gas beyond the inner edge of its evacuated annulus, while the outer planet gives energy and angular momentum to the dust and gas beyond the outer edge of its evacuated annulus. The inner planet gives the outer planet the amounts of energy and angular momentum required for them to stay in mean-motion resonance. When a planet trades energy and angular momentum with the disc on its own, it always loses out and ends up drifting inwards. But when two planets work in partnership, they aren't necessarily losers, and they may move very slowly outwards, or very slowly inwards.

The young solar system

It is thought that the young Saturn bumped into the young Jupiter in the way we have described and entered a 2 : 3 mean-motion resonance with Jupiter (three Jupiter years taking as long as two Saturn years) and then the two planets working in partnership pretty much stopped drifting inwards. Then the next planet out, we'll call it ice giant 1, drifting inwards encountered the almost stationary Jupiter–Saturn pair and entered a mean-motion resonance, probably again 2 : 3 with Saturn, and the system of three locked planets drifted only very slowly in radius. So along comes the next planet out, ice giant 2, and enters a mean-motion resonance with ice giant 1. This resonance may have been 3 : 4. Now all four planets, working in partnership, remained in pretty much the same places while the young Sun dispersed the gas in the disc.

Since the orbital time around a star increases with radius r as $r^{3/2}$, the time taken for dust to accumulate into asteroids and asteroids to gather into planets increases as we move outwards, and beyond $\sim 20\,\mathrm{AU}$ this process was still incomplete when the Sun had dispersed the gas. So there was no ice giant 3 beyond ice giant 2, only a large number $\sim 1{,}000$ of Pluto-sized objects and zillions of asteroids. Once the gas was gone, there was nothing to damp the eccentricities of the asteroids that were excited by the gravitational fields of the Pluto-sized bodies. This ensemble of asteroids and Plutos surrounded the four locked planets but did not extend into the evacuated annulus of ice giant 2. The present-day Kuiper Belt of asteroids is the descendent of this ensemble and we shall refer to the ensemble as the *Primordial Kuiper Belt (PKB)*.

One of the ice giants, probably ice giant 1, was on a slightly eccentric orbit and was able to exchange energy and angular momentum with the asteroids near the inner edge of the PKB at $r < 20\,\mathrm{AU}$. If it wasn't locked in mean-motion resonances to the other planets, it would respond to this loss by moving to a fairly circular orbit of smaller radius. But it is locked, so it responds by moving to a more eccentric orbit. For a while its eccentricity steadily grows and then suddenly a secular resonance within the four-planet system causes the eccentricity of ice giant 1 to decrease, and the angular momenta of the planets to change in such a way that the mean-motion resonant conditions are broken.

With the resonant conditions broken the planets can no longer exchange energy so any loss of angular momentum will lead to an increase in eccentricity (see equation (5.1)). The eccentricities of the two ice giants quickly grow to large values so each of these planets crosses the other's orbit and possibly even Saturn's orbit. This is a time of great peril for the solar system, for a planet on a highly eccentric orbit is likely to induce other planets to move to eccentric orbits, and once Jupiter was on a highly eccentric orbit it would not be long before Jupiter would have driven every other planet either into the Sun or completely out of the solar system.

We shall see below that a catastrophe of this type has probably occurred in many planetary systems. We owe our existence to good fortune and the way the PKB acted as a fire bucket. As the eccentricities of the ice giants grew, they penetrated into the PKB and started to have close encounters with asteroids and Plutos. Scattering objects in the PKB damped the eccentricity of the ice giants, and the system settled to its present configuration. Neptune is now in a 1 : 2 mean-motion resonance with Uranus and on an orbit of low eccentricity and semi-major axis $a = 30.1\,\text{AU}$ that places it far into the PKB. Even the orbit of Uranus probably lies within the PKB (Table 1). In some simulations of the evolution of the four-planet system after the resonance condition is broken, ice giant 1 ends up on a smaller orbit than ice giant two, and in other simulations it ends on the larger orbit. Thus we do not know which ice giant Neptune is.

The population of the PKB was decimated when the ice giants swept through it, so the present Kuiper belt contains only $\sim 0.07\,\mathcal{M}_\oplus$ rather than the $\sim 40\,\mathcal{M}_\oplus$ from which we believe it started, and all but one of the ~ 1000 Plutos and many of the asteroids have been turfed out of the solar system. However, many of these objects at some stage appeared within the PKB and were scattered by the planets from Mercury to Saturn, pitting their surfaces and damping their eccentricities. Indeed, the rate at which asteroids hit the moon can be determined from the pattern of craters they made, and long before our current picture of the evolution of the solar system emerged it was known that there was a *late heavy bombardment* (*LHB*) of the Moon approximately 0.7 Gyr after the formation of the Sun 4.6 Gyr ago. Another likely legacy of this period of high asteroid density are the *Trojan asteroids* of Jupiter, which move on the same orbit as Jupiter but on the other side of the Sun. It is thought that Jupiter captured these asteroids at this time.

Order, chaos, and disaster

Newton bequeathed us a wonderful tool for predicting the future: the differential equation (page 2). Use the numbers that describe the current configuration of a dynamical system as the initial conditions from which you solve the equation, and from the solution you can read off a prediction for the system's configuration at any time. Unfortunately in the 20th century it emerged that this scheme often doesn't work. The problem is that the behaviour of the solution can be *extremely* sensitive to the initial conditions. Henri Poincaré, a cousin of the man who led France in much of 1914–18 war, first sighted this problem at the beginning of the 20th century, but the scale of the problem only emerged in the 1960s as electronic computers made it feasible to solve the differential equations of generic dynamical systems. The solar system provides at once the cleanest and the most awe-inspiring example of this phenomenon.

The differential equations that govern the motion of the planets are easily written down, and astronomical observations furnish the initial conditions to great precision. But with this precision we can predict the configuration of the planets only up to ~ 40 Myr into the future—if the initial conditions are varied within the observational uncertainties, the predictions for 50 or 60 Myr later differ quite significantly. If you want to obtain predictions for 60 Myr that are comparable in precision to those we have for

40 Myr in the future, you require initial conditions that are 100 times more precise: for example, you require the current positions of the planets to within an error of 15 m. If you want comparable predictions 60.15 Myr in the future, you have to know the current positions to within 15 mm. We really are up against a hard limit to our knowledge here in the sense that it is inconceivable that we will be able to specify the current configuration of the solar system to be able to predict the positions of the planets more than ~ 60 Myr in the future. This is a disappointingly small fraction of the $4,600$ Myr age of the system.

In these circumstances all we can do is use the differential equations to compute which configurations are most likely in the future. We do this by randomly sampling current configurations that are consistent with the observational data, and for each sampled configuration computing the corresponding solution so we can predict where the planets will be in that case. The most probable future configurations are those that are nearly reached from many sampled current configurations.

An important feature of the solutions to the differential equations of the solar system is that after some variable, say the eccentricity of Mercury's orbit, has fluctuated in a narrow range for millions of years, it will suddenly shift to a completely different range. This behaviour reflects the importance of resonances for the dynamics of the system: at some moment a resonant condition becomes satisfied and the flow of energy within the system changes because a small disturbance can accumulate over thousands or millions of cycles into a large effect. If we start the integrations from a configuration that differs ever so little from the previous configuration, the resonant condition will fail to be satisfied, or be satisfied much earlier or later, and the solutions will look quite different.

The importance of resonances leads to astonishing sensitivity to the small print of physical law. In Chapter 6 we will introduce

Einstein's theory of relativity and discuss some of its astrophysical applications. Here we anticipate the fact that the importance of relativity is quantified by v^2/c^2, where v is a typical velocity and c is the speed of light. For the Earth this ratio is $\sim 10^{-8}$, so extremely small, and even for Mercury it's less than 2.5 times bigger. Yet an experiment conducted by Jacques Lasker and Mickael Gastineau indicates that we probably wouldn't be here if it weren't for the tiny effect that Einstein's correction to Newton has on the solar system.

Laskar and Gastineau evolved the solar system forward from its present configuration with two ensembles of solutions. In each case they used as initial conditions for the solutions configurations for the solar system that are consistent with the best observational data. They computed one set of solutions 5 Gyr into the future using Newton's theory without Einstein's tiny corrections, and the other set using Einstein's corrections. They found that Mercury attained an eccentricity $e > 0.7$ in just 1 per cent of the solutions when Einstein's corrections were included, whereas without Einstein's corrections only 1 solution out of the 2,500 computed kept Mercury on an orbit with $e < 0.7$. If Mercury does attain $e > 0.7$, the consequences for the Earth are dire, because an eccentric Mercury soon excites high eccentricity in Venus, which drives the Earth to high eccentricity, which drives Mars to high eccentricity. Various dramatic consequences ensue: in the 2,500 solutions to Newton's uncorrected equations, Mercury collided with Venus in eighty-six cases, and collided with the Earth in thirty-four cases. Even if the inner planets don't collide with one another, they are either driven into the Sun or flung right out of the solar system.

So it seems we live on the edge of a precipice, and were it not for Einstein's absolutely tiny corrections to Newton's equations, planet Earth would almost certainly not be in a position to offer us shelter. Even with Einstein's corrections, our security is not guaranteed longer than ~ 80 Myr into the future. Nevertheless, thanks to general relativity planet Earth is overwhelmingly likely to survive until it is engulfed by the swelling Sun $\sim 4\,\text{Gyr}$ from now.

Einstein's corrections make life possible by de-tuning a resonance between Jupiter and Mercury. Because this is a weak resonance, it can have an impact only if conditions are just right. Einstein's corrections make it hard for them to be right.

Extra-solar systems

The first unambiguous detection of extra-solar planetary system was made a recently as 1995 by Michel Mayor and Didier Queloz, but now over a thousand planetary systems are known. We can gain insight into the evolution of planetary systems by considering the statistics of the configurations of the known systems: it's as if the Universe has computed a large number of solutions to evolutionary equations that are guaranteed absolutely correct.

We have seen that the solar system has retained eight planets on nearly circular orbits in the teeth of at least two grave threats: first about 700 Myr into its life when the tight formation of the giant planets was destabilized, and even today the stability of the inner solar system would be imperilled by changing its physics by a part in 100 million.

So it is not surprising that the first systems to be discovered resemble the endpoint of one of these catastrophes in that they consist of a Jupiter-like planet on a fairly short-period, eccentric orbit. However, the significance of this finding must be tempered by the fact that the observational technique used to find the first systems strongly favoured finding systems with a large planet on a short-period orbit: the systems were found by monitoring the velocities of stars for the periodic changes in velocity that signal motion of the star around the centre of mass of the star and its planet. The more massive and close the planet, the larger these velocity changes are and therefore the greater the likelihood that they will be detected above the noise.

Subsequently, a completely different technique has been used to detect planets, and with this technique systems with several planets on nearly circular orbits can be, and have been, found. The technique involves monitoring the brightnesses of stars to detect the slight drop in brightness as a planet passes between its star and Earth. The only systems that can be detected with this technique are those we view almost exactly edge-on, and data of the required precision can only be gathered from space. In May 2009, NASA launched the Kepler satellite to monitor stars in a small area of the sky. In the following four years Kepler definitely detected nearly a thousand planetary systems and drew up a list of several thousand stars that showed signs of planetary systems. Assessing the significance of these data is a very active field of research.

Chapter 6
Relativistic astrophysics

The familiar world of Newtonian physics is an approximation to relativistic physics, which is convenient and works well when relative velocities are significantly smaller than the velocity of light. Astronomers have identified many kinds of object that violate this restriction, and we have to use the full theory of relativity to model these objects.

We will outline the main results of relativity theory later, in 'Special relativity', but we need the most famous result now: $E = mc^2$. This equation summarizes the requirement that according to relativity, a particle's energy does not consist only of its kinetic and potential energies, as in Newtonian physics, but includes in addition its *rest-mass energy* $E_0 = m_0 c^2$, where m_0 is the mass the particle has when it is at rest. It is a number that is characteristic of the particle and never changes. By contrast the mass $m = E/c^2$ changes as work is done on or by the particle: when an electron is accelerated in a collider such as the Stanford Linear Collider (SLAC), its mass increases by a factor ~ 200. The ratio $\gamma = m/m_0$, which is called the *Lorentz factor*, quantifies how relativistic a particle is. A car moving at 100 km/h $\simeq$ 60 mph, has a Lorentz factor that differs from one by a tiny amount ($\gamma - 1 = 5.8 \times 10^{-13}$) so it is distinctly non-relativistic. Typically, any body that has $\gamma - 1 \gtrsim 0.1$ is considered to be moving relativistically.

We now list some situations in which we need to use relativistic theory.

Radio galaxies   Much of the radiation we detect from a radio galaxy is generated by electrons that have Lorentz factors $\gamma \sim 10^5$. The motion of these hyper-energetic electrons is mostly random, but in the core of a radio galaxy there is often a jet in which there is a systematic flow of plasma in which the bulk kinetic energy of the plasma is several times larger than its rest-mass energy.

Micro-quasars   These objects are essentially scaled down versions of radio galaxies: a black hole that drives a micro-quasar has a mass of a few solar masses (page 68) while a black hole in a radio galaxy might have a mass of a few hundred million solar masses. Scaling down the mass makes the phenomena physically smaller and more rapidly fluctuating, but does not change the characteristic velocities. Hence relativity is as important for a micro-quasar as for a radio galaxy.

Gamma rays   Positrons, the anti-particles of electrons, are an essentially relativistic phenomenon—P.A.M. Dirac predicted their existence while trying to make quantum mechanics consistent with relativity. When an electron annihilates with a positron, all the energy of the two particles is converted into photons, usually two photons. If the electrons are non-relativistic (so $\gamma \simeq 1$) each photon has the rest-mass energy of an electron, 511 keV (kilo-electron-volt). Gamma-ray telescopes have detected this spectral line from the direction of the centre of our Galaxy, implying the existence of a significant density of positrons there.

In 1963 the UK, the USA, and the Soviet Union signed a treaty banning tests in the atmosphere of nuclear devices. Neither side trusted the other and the USA and the USSR launched top-secret satellites that would detect gamma rays emitted by elicit tests. To everyone's surprise *many* bursts of gamma rays were detected.

The bursts lasted from seconds to a minute, and they occurred too often to be plausibly generated by nuclear devices.

After the military experts on both sides had puzzled over the data in secret, each side learnt that the other saw these events, and it became clear that the sources were astronomical. In 1973 the data were made public and it was the turn of the astronomers to be puzzled. The events seemed to be uniformly distributed over the sky, which indicated that their sources were not associated with stars in our Galaxy as most X-ray sources are. The sources had to be either within $\sim 0.1\,\text{kpc}$ of the Sun or spread through a volume much bigger than our Galaxy. But the timescales of the sources were much too short for them to be associated with active galaxies, and nobody could come up with a credible source close to the Sun. In 1986 Bohdan Paczynski had the courage to posit that, despite their small timescales, they *are* at cosmological distances, and probably associated with some kind of exploding star. In 1997 this conjecture was proved correct when the William Herschel telescope took photographs of the region around a burst that had just been detected, and the rapidly fading optical *after-glow* of the event was seen in a distant galaxy. Since then optical after-glows have been routinely detected, and we have optical spectra of the underlying objects. These data establish that many gamma-ray bursts are indeed associated with exploding stars. It has also emerged that there is more than one kind of source of gamma-ray bursts, and our understanding of these objects is incomplete. What is certain is that relativity plays an essential role in understanding these extraordinary objects.

Cosmic rays The Earth is constantly bombarded by relativistic particles—the detection of such particles is the oldest branch of high-energy astronomy. Fortunately for us, most of these dangerous particles collide with an oxygen or nitrogen nucleus high in the atmosphere. That nucleus is badly damaged by the impact and shards flash downwards, smashing into other nuclei as they go. So a single energetic particle entering the atmosphere creates a cosmic-ray *shower*.

The particles hitting the Earth have a variety of energies but there are many more low-energy particles than high-energy ones. The most energetic particles are seen only rarely even by the detectors that have the largest collecting areas. Nonetheless, the most energetic particles so far detected have $E \sim 10^{20}$ eV, so if they are electrons they have $\gamma \sim 10^{14}$ and if they are protons or neutrons they have $\gamma \sim 10^{11}$. These energies are way higher than those achieved by the most powerful particle accelerators: the Large Hadron Collider in Geneva currently accelerates protons to $\gamma \simeq 10^3$.

Neutron stars The escape speed from the surface of a neutron star is only $\sim c/2$, so these objects are only mildly relativistic. But they lend themselves to precision measurements, so the modest relativistic effects can be precisely quantified and provide strong tests of relativity theory. Of particular importance is the *Hulse-Taylor* binary pulsar PSR B1913+16, a pair of neutron stars in orbit around one another with eccentricity $e = 0.62$ and period of 7.75 hours. This binary was discovered by Joe Taylor and his graduate student Russell Hulse in 1974, and has been intensively observed since that date. A few similar binary neutron stars have been discovered since, but the constraints they place on relativity theory are weaker because they have not been observed for so long.

X-ray sources We saw in Chapter 4, 'Accretion discs', that the innermost radii of the accretion discs around black holes are so hot that they radiate most strongly in X-rays. These regions are too tiny to be resolved by X-ray telescopes, but relativistic effects modify the shapes of X-ray emission lines that we detect.

The solar system The earth orbits the Sun at a speed of ~ 30 km s^{-1}. Since this speed is $\sim 10^{-4}c$ and relativistic effects tend to be smaller than Newtonian ones by a factor $\sim (v/c)^2$, you would expect relativity to have a very small impact on the solar system. Nonetheless, since the solar system lends itself to precision measurement, the data require relativity for their interpretation and provide crucial tests of the theory of relativity. Moreover, as we saw in Chapter 5, 'Extra-solar systems', the dynamics of the solar system is delicate and if relativistic effects were absent, the configuration of the solar system would be qualitatively different from what we actually see.

The Universe   Conceptual problems make it impossible to develop a persuasive model of the dynamics of the Universe within Newtonian physics. Hence cosmology was opened up as a branch of physics by Einstein's theory of general relativity. In the 1960s cosmology was put on a solid empirical basis by the discovery of the *microwave background radiation*, which allows us to study the Universe as it was only 100,000 years after the Big Bang, and the discovery of quasars, most of which are in galaxies that are receding from us at relativistic speeds.

Special relativity

On page 4 I explained that physicists are determined that the laws of nature shall be the same everywhere and at all times—any change in the measured phenomena from place to place or from time to time *must* be traced back to some change in the conditions under which the universal laws should be applied. The special theory of relativity articulates a new requirement for invariance: the laws shall be same in all galaxies, no matter how fast they move with respect to one another, and in all spaceships, no matter how speedy.

In 1899 Hendrik Lorentz discovered a symmetry in Maxwell's equations of electrodynamics. We now call this symmetry *Lorentz covariance* and consider it to be a fundamental principle of physics, but Lorentz was unclear what the physical significance of this new symmetry was. In 1905 Einstein argued that Lorentz's symmetry reflected the fact that electromagnetism works in exactly the same way in any spaceship, regardless of the spaceship's velocity. This assertion surprised Einstein's contemporaries because Maxwell's electromagnetic waves must be propagating through some medium, the *aether*, and the aether could not be at rest with respect to all spaceships. Indeed, the velocity of spaceship Earth changes by ~ 60 km s^{-1} every six months, and experiments with light should be able to detect such a change. Einstein's contemporaries were puzzled that no such experiment had yet succeeded in detecting our motion with respect to the aether. Einstein argued that on account of Lorentz's symmetry, it is in principle impossible to detect motion with respect to the aether. This medium, which we now call the *vacuum* has the remarkable property of looking the same to all observers, no matter how they move with respect to one another. In particular, no observer can be said to be absolutely at rest, so all motion is relative to some other observer—hence the name of Einstein's theory.

The key to getting to grips with relativity is to analyse every physical situation as a series of *events*. An event happens at some place *and some time*. So an event is specified by four numbers, the x, y, and z coordinates of its place, and the time, t, at which it occurred. An observer O' who moves with respect to the observer O who uses the four numbers (x, y, z, t) will assign a different set of numbers (x', y', z', t') to the same event. A simple rule, the *Lorentz transformation* discovered by Lorentz, enables one to calculate the primed numbers from the unprimed numbers given the velocity **v** at which O' moves with respect to O.

The extraordinary thing about a Lorentz transformation is how it treats two events (x, y, z, t) and (x_1, y_1, z_1, t_1) that are simultaneous for O, so $t_1 = t$: in general they are *not* simultaneous for O' in the sense that $t' \neq t'_1$. Hence O' considers that one event happened before the other even though O knows for a fact that they occurred simultaneously. This idea that simultaneity is observer-specific is very counter-intuitive and hard to get used to.

It underlies an absolutely bewildering principle: moving clocks run slow. For example, Bob standing on a station platform studies carefully the clock on the laptop of Alice on an express as the train whizzes by. He concludes that the clock is running slow because it's a moving clock. Meanwhile Alice studies Bob's watch and concludes that his watch is running slow because it's moving with respect to her. Actually the clock is keeping perfect time when checked by Alice, and the watch likewise runs perfectly in Bob's eyes, but both run slow when checked by a moving observer. The factor by which the timepieces run slow is the Lorentz factor γ associated with the train's velocity.

Muon lifetimes

Cosmic rays detected on Earth provide a direct confirmation of the moving-clocks principle. Muons are elementary particles akin to electrons that are highly unstable so they rapidly decay into other particles—their *half-life* (the time required for half of a sample to decay) is 2.2 μs. Muons are created when cosmic ray particles hit an atomic nucleus $\sim$ 20 km up in the Earth's atmosphere. Even moving at the speed of light, they can only travel $\sim$ 660 m in a half-life. Hence you might think that very few, if any, of the muons created 20 km above the ground would reach the ground, contrary to what was found by sending detectors up in balloons. Relativity resolves this paradox by stating that while our clock moves through 2.2 μs, the muon's own clock advances by $2.2/\gamma$ μs, so our clock must advance by 2.2γ μs before the muon decays, and a muon that moves at speed $\sim c$ travels 660γ m before it decays. Hence muons created in the upper atmosphere with $\gamma \gg 1$ have a good chance of reaching the Earth.

Rest-mass energy

Alice and Bob won't agree on the energy or momentum of a given particle, because if the particle is stationary in Alice's frame, Alice will say it has no kinetic energy and no momentum, while Bob will say it has both kinetic energy and momentum. So there has to be some rule for deducing the energy and momentum that Bob assigns from the values assigned by Alice. Wonderfully, this rule turns out to be a Lorentz transformation. So if (p_x, p_y, p_z, E) are the components of momentum and the energy of a given particle of mass m_0 as seen by Alice, then we can compute the momentum and energy (p'_x, p'_y, p'_z, E') seen by Bob by applying the Lorentz transformation for the velocity of Bob with respect to Alice to the four numbers $(p_x, p_y, p_z, E/c^2)$ rather than the coordinates (x, y, z, t) of an event. This simple rule implies that the energy of a particle of rest mass m_0 is $E = \gamma m_0 c^2$. In particular, when $v \ll c$ and $\gamma \simeq 1$, we $E = m_0 c^2$, Einstein's famous expression of the equivalence of mass and energy.

According to quantum mechanics, a photon of angular frequency ω, wavelength λ that moves along the unit vector **n** has energy $\hbar\omega$, and momentum $\hbar\mathbf{k}$, where $\hbar$ is Planck's constant divided by 2π and **k** is the *wavevector* $\mathbf{k} = (\omega/c)\mathbf{n}$. When we apply a Lorentz transformation to the four numbers (k_x, k_y, k_z, ω) used by Alice, we obtain the wavevector $\mathbf{k}'$ and angular frequency ω' of the photon in the eyes of Bob. Since the photon is really a train of waves, we expect the frequency ω' measured by Bob to be *Doppler-shifted* with respect to that ω measured by Alice, and the Lorentz transformation provides the right way to compute the shift. It also enables us to discover how the direction in which the photon travels is affected by the motion of Bob with respect to Alice because these directions are given by the unit vectors $\mathbf{n} = (c/\omega)\mathbf{k}$ and $\mathbf{n}' = (c/\omega')\mathbf{k}'$.

Thus different observers disagree about the directions in which a given photon moves. An example should make it physically obvious that there will be disagreement. In an old-style western movie a guard on a moving train that is carrying gold is firing his rifle at bandits. If the guard fires perpendicular to the car, the bullet will not move perpendicular to the track because in addition to its velocity down the barrel it shares in the train's forward motion. So if the guard wants to hit a target that lies at right angles to the track at his present location, he needs to aim backwards from perpendicular to the car, so there is a component of the bullet's velocity down the barrel that cancels the forward motion of the train.

Now imagine a panicked posse of guards on the train who spray bullets uniformly in all directions. Then half of their bullets will be fired in directions forward of perpendicular to the car, and half backward of perpendicular to the car. Consequently, viewed from the ground more than half the bullets will be forward moving: the motion of the train causes the shower of bullets to be *forward beamed*.

Jets

As we saw in Chapter 4, 'Jets', several astronomical objects emit jets of plasma. The bulk velocity of the jet may yield a Lorentz factor γ of several. Within the jet photons are emitted by a variety of processes with an angular distribution that is roughly uniformly distribute in angle when viewed by an observer travelling with the material in the jet. From our perspective, these photons, like the posse's bullets, emerge with a strong bias to the forward direction—the emission is forward beamed. In fact half of the emitted photons will be travelling in directions around the jet axis that occupy just $1/(2\gamma^2)$ of a sphere. Even for modest values of γ, this is a very small fraction of the sphere. More generally Figure 26 shows the density of photons on the sphere as a function of the angle θ between the photon's direction and the jet axis. For negligible jet velocity ($\gamma \simeq 1$) the density is one for all θ. We see that even for $\gamma < 2$ forward beaming is a strong effect. In fact, the forward concentration of the *energy* emitted by the jet is even larger than Figure 26 would suggest because the photons received in the forward direction have their frequencies, and thus their energies, most strongly raised by the Doppler effect.

In general we expect an object that's prone to jet formation to emit a pair of oppositely directed jets. Generally one jet will have a

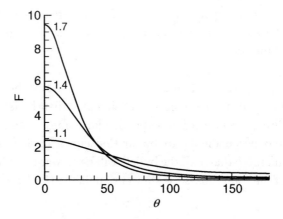

26. The angular distribution of photons emitted by jets with Lorentz factors γ = 1.1, 1.4, and 1.7. The quantity plotted is the the number of photons per unit solid angle at the given angle. For an isotropic distribution this is unity.

component of velocity towards us, and the other away from us. On account of forward beaming, the approaching jet will be brighter than the receding one. Since an object has to exceed a critical brightness to be detected at all, we may detect only the approaching jet. This possibility is likely if we happen to view the object from close to the line along which the jets are being fired. Many radio galaxies appear to have only one jet.

Jets sometimes display *superluminal motion*, which is forbidden by relativity. The phenomenon is for blobs to be seen within a jet that move across the sky at a speed that seems to exceed the speed of light. The speed in question is evaluated by assuming that the blobs are moving within the plane of the sky on the ground that any component of velocity perpendicular to the plane of the sky will only increase the distance travelled, and therefore the speed derived.

Figure 27 shows the relevant geometry. A single blob is shown by an open circle twice, once at time t' (upper circle) when the blob is close to the source, and at a later time t when it is further from the

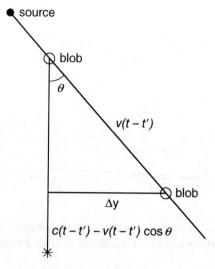

27. The geometry of superluminal expansion.

source. At both t' and t the blob emits photons towards Earth and the location at time t of the first of these photons is shown by a star at the bottom. This photon has a head start on the second photon in the race to Earth by distance $c(t - t') - v(t - t')\cos\theta$ so it will arrive at Earth sooner by

$$\Delta t = (t - t')\left(1 - \frac{v}{c}\cos\theta\right).$$

Between emitting the two photons the blob has moved a distance $\Delta y = v(t - t')\sin\theta$ over the sky. So the blob's apparent speed is

$$v_{\text{app}} = \frac{\Delta y}{\Delta t} = \frac{v\sin\theta}{1 - (v/c)\cos\theta}.$$

Figure 28 shows v_{app} as a function of θ for three values of v/c: 0.2, 0.71, and 0.96. It shows that superluminal expansion is possible for $v > 0.71c$ and that for $v \gtrsim 0.95c$, v_{app} can be several times c. Superluminal motion has been observed in many radio sources.

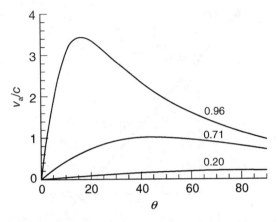

28. The apparent velocity of a blob that moves at speed $v = 0.96c$ or $0.71c$ or $0.20c$ along a line that is inclined at angle θ to the line of sight.

Shocks and particle acceleration

The space surrounding a relativistic object that fires jets is never completely empty. As the jet emerges from the source, its density may be so high that the density of the ambient medium is negligible, but as the jet moves away from the source it spreads laterally and its density declines, so eventually the ambient medium will have a significant impact on the jet.

The easiest way to imagine what happens when a jet ploughs through the ambient medium is to imagine you are moving at a velocity that is intermediate between that of the jet and the ambient medium. Let's say the jet is approaching from your left and the ambient medium is approaching from your right. The space where you stand is filled by material that was produced by the collision of jet material on ambient material. This material is incredibly hot because the ordered kinetic energy of the jet coming from the left and of the ambient material coming from the right has been randomized into frenetic darting here and there by individual particles: individual particles are zipping along, but on the average they are going nowhere, so this *shocked plasma* is at rest with respect to you.

The region of shocked plasma moves away from the source and grows steadily as a result of fresh jet material hitting it from the left and ambient material hitting it from the right. The narrow regions in which the systematic motions of the jet and ambient particles are randomized are called *shocks*.

Typically the shocks are what physicists call *collisionless* shocks. Given that a shock is precisely a place where fast-moving fluid collides with slower fluid, the collisionless sobriquet sounds daft. What it means is that the velocities of particles are changed by an electromagnetic field that has a much longer lengthscale than the inter-particle separation. Thus an incoming electron or proton is not decelerated by colliding with an individual atom or ion, but by a fairly smooth electromagnetic field generated by zillions of electrons and ions collectively (Figure 29).

The origin of this field is fundamentally separation of electrons from ions arising from the enormous mass difference (by a factor in excess of 1,800) between electrons and ions: incoming electrons decelerate much sooner than the more massive ions, so regions of

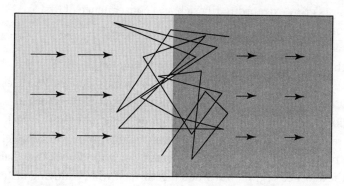

29. Plasma coming fast from the left hits slower-moving plasma on the right. On impact the plasma slows down and becomes denser, symbolized the by the smaller arrows and darker shading. The zig-zag line shows the trajectory of a fast particle that is multiply scattered, often crossing the interface between the two regions between scatterings.

positive and negative charge density arise. These regions generate an electric field which pulls one way on the electrons and the opposite way on the ions, thus tending to bring them to the same mean velocity. Differences in the mean velocities of electrons and ions imply the existence of electric currents, and these currents generate a magnetic field. Moreover, the flows of electrons and ions is highly unsteady in this region, so the electric and magnetic fields are time dependent. Because the fields are time-dependent, they can change the energies of individual electrons and ions, and on the average they transfer energy from the ions to the electrons – in the original, ordered flow upstream of the shock, both species had the same velocity, so the kinetic energy of the material was overwhelmingly contained in the ions. The post-shock plasma is *relaxing* to a state of thermal equilibrium in which each species has precisely half the now randomized kinetic energy. Net transfer of kinetic energy from the ions to the electrons is a key part of this relaxation process.

The shocked plasma is analogous to a gambling den into which punters bring money which is redistributed in the den. We've just discussed how this transfer affects the average punter. But a few punters grow extraordinarily rich as a consequence of first becoming unusually rich.

In the plasma the analogue of wealth is kinetic energy, and the faster a particle moves, the harder it becomes to deflect. Particles that become sufficiently fast can crash right out of the shocked region, and thus enter one of the regions to right or left in which there is an ordered inflow. Since these regions are very extensive, the particle *will* eventually be deflected there and find its way back to the shocked region. But when it returns it will be moving faster than when it departed because the net effect of its deflections in the ordered flow is to reverse the sign of its velocity *with respect to the flow*. In the non-relativistic case its speed is now the sum of its original speed and the speed of the inflowing material. Since the particle is now moving faster than ever, it is likely to crash right

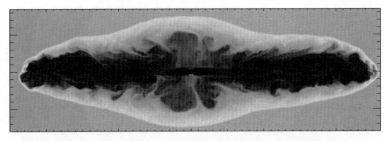

30. Computer simulation of a jet-inflated cocoon. Dark shading indicates low density. The grey at the edge represents the undisturbed circumgalactic gas. The light band inside this is the result of compressing this gas in a shock. The dark shades further in show extremely hot, turbulent plasma that was heated in the shocks at the ends of the jets.

through the shocked region and enter the opposite region of ordered inflow, where again the sign of its velocity with respect to that flow will be reversed and it will return at an even greater speed. By this process of *Fermi acceleration* particles can acquire very large Lorentz factors. In fact this is how the cosmic rays we detect on Earth are accelerated.

Because it is very hot, the shocked plasma is a high-pressure fluid and expands in any direction it can. The *ram pressure* of the jet to the left and of the ambient medium to the right prevent it expanding in these directions, but it can usually expand in the perpendicular directions. Flowing outwards perpendicular to the jet in this way it gradually inflates a *cocoon* of plasma that surrounds the jet as sketched in Figure 30.

Synchrotron radiation

Within the cocoon there are many electrons that were accelerated to large Lorentz factors by the Fermi process. The cocoon invariably contains a magnetic field, and the electrons spiral around the field's lines of force, emitting electromagnetic radiation as they do so. If the electrons are non-relativistic, the radiation is all at one frequency, the *Larmor frequency* v_L, which is proportional to the strength of the magnetic field. If the electrons are relativistic, the radiation covers a band of frequencies that extends up to $\gamma^2 \nu_L$ and the radiation is called *synchrotron radiation*. The typical Lorentz factor of the emitting electrons can be inferred from the spectrum of the radiation. Because magnetic fields in interstellar and intergalactic space are generally weak, radio telescopes are only sensitive to synchrotron radiation from electrons with Lorentz factors $\gamma \gtrsim 10^4$, yet sychrotron radiation is often observed.

General relativity

We have seen that the special theory of relativity emerged from Maxwell's equations of electrodynamics, but it laid bare a fundamental symmetry of space-time. Einstein became convinced that *all* fundamental physical laws should display this symmetry, and a theory that conspicuously failed to do so was Newton's theory of gravity.

There is an extremely close analogy between gravity and electrostatics: like gravity, the electrostatic force is proportional to the inverse square of distance. The theory of relativity revealed that magnetism is a relativistic correction to electrostatics in the sense that a moving observer sees an electric field in part as a magnetic field, so we expect a gravitational field to look rather different to a moving observer. In particular, we should expect the complete gravitational force on a body to depend on its velocity, just as the electromagnetic force on a charge has a magnetic component that's proportional to velocity. Since photons move faster than any particle of non-zero rest mass, understanding this component must be essential if you want to understand how a gravitational field affects photons.

There's a special aspect to gravity that electrostatics lacks and Einstein was convinced was of fundamental importance. This is that the gravitational force is proportional to mass. Legend has it that Galileo demonstrated this fact around 1600 by dropping two balls with very different weights from the top of the Leaning Tower of Pisa: despite their different weights, the balls landed essentially simultaneously. Galileo's experiment was not very precise. A much better test of the dependence of the gravitational force on mass is to measure the periods of pendulums that have the same length but bobs made of different materials and masses. In 1891 Baron Roland Eötvös devised an experiment that demonstrated the proportionality to extremely high precision by probing the balance between an object's gravitational attraction to the Sun with the force required to keep it in orbit around the Sun. Einstein considered that such a precise proportionality couldn't happen by chance; it must emerge as inevitable from the correct theory of gravity.

Einstein struggled for ten years to put mathematical flesh on this idea, and the theory he produced is undoubtedly one of the greatest creative achievements of mankind. In essence, he did for gravity what Maxwell had done for electromagnetism, namely draw disparate bits of physics into a single coherent mathematical structure, which contained not only the physics that inspired it but also predictions of entirely new phenomena.

An electromagnetic field is generated by the density of electric charge and current. A gravitational field is generated by the density of energy-momentum and the fluxes of this quantity. In Newtonian physics gravity is generated by mass, which through $E = mc^2$ is equivalent to energy. General relativity teaches that this view is too limited, and arises because we haven't had experience of fast-moving massive bodies or very high pressures, so we are unaware that both momentum and a flux of energy-momentum also help to generate the gravitational field. We also think of gravity as an intrinsically attractive force, while general relativity shows that it can equally be repulsive.

Maxwell's equations are differential equations that can be solved for the electromagnetic field generated by a given density of electric charge and current, and Einstein's equations are differential equations that can be solved for the gravitational field that is generated by a given density and flux of energy-momentum. Whereas Maxwell's equations are linear, Einstein's equations are *non-linear equations*: if an equation is *linear*, you can find simple solutions and add them together to build more complex solutions. When an equation is non-linear, the sum of two solutions is generally not a solution, so you cannot build up sophisticated solutions by adding simple ones.

On account of this problem the only exact solutions of Einstein's equations that we have are ones with special symmetries. The first and most famous of these solutions is that found in 1916 by Karl Schwarzschild. This solution describes the gravitational field that surrounds a spherical mass, and in Chapter 5, 'Extra-solar systems', we discussed its application to the solar system. In 1963 Roy Kerr extended this solution to the gravitational field that surrounds a spinning body, and this solution is important for understanding the inner regions of accretion discs (Chapter 4, 'Journey's end'). Other exact solutions describe homogeneous universes. A exact solution from this rather limited set can be extended to an approximate solution by perturbation theory (cf. Chapter 5, 'Dynamics of planetary systems'), and many astronomical applications of general relativity rely on this approach. Since 2000 critical advances have been achieved in the numerical solution of Einstein's equations. These advances arise in part from the steady growth in available computing power, but they are mainly due to better understanding of how the equations need to be tackled.

Weak-field gravity Two conditions have to be satisfied for Newton's theory of gravity to be accurate: first, the gravitational field must be weak and speeds must be much less than c. The second condition is the more restrictive in practice, not least because photons always violate it. In astronomy the field is usually weak and time-independent. Then the gravitational field is completely described by the Newtonian gravitational potential Φ—for example, distance r from a point mass M we have, $\Phi(r) = -GM/r$.

Gravitational redshift

Suppose we measure the frequency of a spectral line emitted by atoms on the surface of a compact object such as a white dwarf. Then standing at $\mathbf{x}_o$ we are effectively listening to the ticks of an atomic clock that does not move relative to us and sits at $\mathbf{x}_c$ near the compact object. Suppose the clock ticks once a second. Then the time that elapses between our receiving light pulses sent out with each tick is

$$t_o = \sqrt{\frac{1 + \tfrac{1}{2}\Phi(\mathbf{x}_o)/c^2}{1 + \tfrac{1}{2}\Phi(\mathbf{x}_c)/c^2}}\ \text{s}.$$

Since the clock is closer to the compact object than we are, $\Phi(\mathbf{x}_c) < \Phi(\mathbf{x}_o)$ and we have $t_o > 1\,\text{s}$. That is, we think the clock is running slow. Reverting to the case in which the clock is an oscillating atom, we have that the measured frequency $\nu_o = 1/t_o$ is lower than the intrinsic frequency $\nu_c = 1/t_c$, where t_c is the time between ticks as measured at $\mathbf{x}_c$. This is the phenomenon of *gravitational redshift*.

Gravitational lensing

Light travels through glass or water slower than it does through air, with the consequence that rays of light are bent when they enter glass or water. This phenomenon is usually quantified by the *refractive index n*, which is defined such that the speed of light in a medium of refractive index n is c/n.

If we use ordinary Cartesian coordinates (x, y, z) and assume that the distance between two points x_1 and x_2 on the x axis is $s = x_2 - x_1$, then a gravitational field endows the vacuum with a refractive index

$$n = 1 + \frac{|\Phi|}{2c^2}. \tag{6.1}$$

With the rather natural definition of distance that we've adopted, when $\Phi \neq 0$ light appears to travel through the vacuum slower than the speed of light. Actually it always travels exactly *at* the speed of light and our expression gives a lower value because we have under-estimated the distance between x_1 and x_2. But it is very useful to imagine that photons travel slower than c when $\Phi(\mathbf{x}) < 0$.

This conceit is helpful because it allows us to use our knowledge of optics to predict how light will be deflected by a gravitational field. A lens brings a parallel beam of light to a focus by slowing photons that pass through the centre of the lens more than it slows photons that pass far from its axis, where it is thinner (Figure 31, upper panel). In fact, a lens is designed such that the time taken by photons to travel from a distant source to the image is

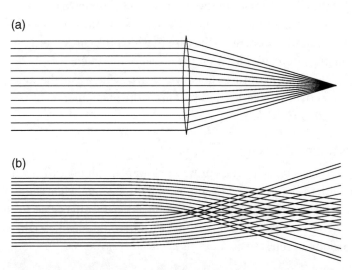

31. A lens is designed to cause a bundle of initially parallel rays to all pass through the focus (a). A gravitational lens (b) causes initially parallel rays to converge but does not cause them to pass through a single point. The mass of the lensing object is unrealistically massive by many powers of ten so the deflections can be easily seen.

independent of the distance from the axis at which a photon passes the lens—this is *Fermat's principle of least time*.

Sometimes our line of sight to a distant object such as a quasar passes close to the centre of a massive object such as a galaxy or a cluster of galaxies. Then the gravitational field of the intervening object acts like a lens in causing rays that were diverging from the source to converge on a point near us. The lower panel of Figure 31 shows a quantitative example of this phenomenon.

As Figure 31 illustrates, the lens formed by a typical gravitational field is of poor quality in that it does not cause all rays to cross at a point, and consequently it produces out-of-focus and distorted images. In fact, a gravitational lens is liable to form several images of the same object. In 1979 the first example of this *strong lensing* phenomenon, SBS 0957+561, was discovered, and it remains one of the most dramatic examples: a galaxy at redshift $z = 0.355$ forms two images of a single quasar 6 arcseconds apart. Since the discovery of SBS 0957+561 systematic searches for multiply imaged quasars have been conducted and hundreds of examples have been found, some with four images, but almost always smaller image separations than that of SBS 0957+561.

As we saw in Chapter 4, 'Time-domain astronomy', the luminosities of quasars fluctuate quite significantly on timescales of months or years. When a quasar is multiply imaged, each image is associated with a particular time of passage of photons from the quasar to Earth. The differences in these times can be measured by finding the time by which one has to shift forwards or back a record of the brightness of one image to make it match the brightness record of another image. These time delays can be computed for any model of the gravitational field that forms the lens, and they are proportional to the *Hubble constant* (H), which relates redshift (z) to distance (s) through $z = Hs/c$, because this determines the distance to an object of known cosmological redshift: the bigger this distance is, the longer is the journey the

32. Weak lensing in the galaxy cluster Abell 2218. The images of galaxies that lie behind the cluster are stretched by the cluster's gravitational field into arcs perpendicular to the field direction.

photons have taken to reach us, and the greater is the time delay associated with choosing a longer route.

Multiple imaging of quasars and galaxies is a rare phenomenon. What's extremely common is for the gravitational field along lines of sight to distant galaxies to distort and magnify their images (Figure 32). This phenomenon is called *weak lensing*. In weak lensing the gravitational field plays the role of badly polished primary lens of a gigantic telescope. Weakly lensed galaxies may be sufficiently brightened by the lens for it to be possible to study them in much greater detail than normal galaxies at their redshift.

Astronomers can even take advantage of the poor quality of gravitational lenses. A weakly lensed image is stretched perpendicular to the direction of the gravitational field. So if we viewed a population of intrinsically round galaxies through a gravitational lens, the galaxies would appear elliptical. The short axes of the ellipses would be parallel to the projection onto the sky of the gravitational field, and the axis ratios of the ellipses would indicate the strength of the field. Using this idea to measure gravitational fields is hard in practice, not least because galaxy images are intrinsically elliptical. However, distortion of images by the gravitational field tends to align the ellipses of galaxies that are close on the sky, and precision measurement of this effect is now a major tool for cosmology.

Gravitational micro-lensing

When a foreground star passes in front of a background star, the gravitational field of the foreground star may strongly lens the background star. Usually the images of the background star form so close together that current telescopes cannot separate them, but lensing can nonetheless be inferred from the way the background star brightens as the gravitational field focuses its light on the Earth. The formation of unresolvable multiple images is called *micro-lensing* and constitutes an important probe of our Galaxy's content. Often the foreground star is too faint to be detected, so all we see is a single star temporarily surging in brightness.

Since the late 1990s the brightnesses of hundreds of millions of stars have been monitored every night and thousands of instances of micro-lensing have been discovered. Interpretation of the data is difficult because many stars have fluctuating luminosities. There are two ways to distinguish luminosity variations from micro-lensing: (i) the former is generally associated with a colour change while micro-lensing is not; and (ii) a single star has a negligible probability to be micro-lensed more than once in recorded history, while a star that fluctuates in luminosity once is very likely to do it often.

Micro-lensing is important because it allows us to detect the gravitational fields of objects that are far too faint to be directly seen. In fact, the probability that any given star is being micro-lensed at any given time depends only on the mass density of the lensing objects along the line of sight to the star, and not on the mass of an individual object. So the probability that a given star is being micro-lensed tonight might be 10^{-6}, regardless of whether the mass density is made up of black holes of mass 1,000 $M_\odot$, or stars of mass 1 $M_\odot$, or Jupiters of mass 10^{-3} $M_\odot$. What changes between these cases is the duration of a typical micro-lensing event, which is proportional to the mass of the lensing object. So if black holes are responsible for lensing, each event will last a million times longer than if Jupiters do the job, and there will be a million times fewer events per year. Hence by monitoring the brightnesses of stars, one can determine both the density of the lensing objects and their typical mass. If the lensing objects are too massive, we'll have to be very lucky to detect a single event, and if their masses are too small the events will be so brief that we won't have enough observations during any given event to detect the characteristic rise and fall in brightness that distinguishes a microlensing event from noise. But the range of masses to which observations are sensitive is very large $\sim 100 - 10^{-3}$ $M_\odot$. Thus microlensing has been used to place upper limits on the space density of very low-mass stars and free-floating planets within our Galaxy. It has also been used to detect planetary systems that could not otherwise be detected.

When a routinely monitored star begins to brighten in a way suggestive of micro-lensing, a model of the lens is fitted to the data. If this model suggests that the background star is going to pass very close to the centre of the lens (where an unseen star sits), observers all round the world, including many amateur astronomers, are alerted because it is then key to keep the star under twenty-four-hour surveillance, and this cannot be achieved from just one or two sites. In the few hours that the star is very close to the centre of the lens, the contributions of planets to the lensing gravitational field can modify the measured brightness quite significantly, and thus betray their existence (Figure 33).

Deflection of light by the Sun

The gravitational field of the Sun forms a lens in which we are embedded. At our distance from the Sun, equation (6.1) gives a

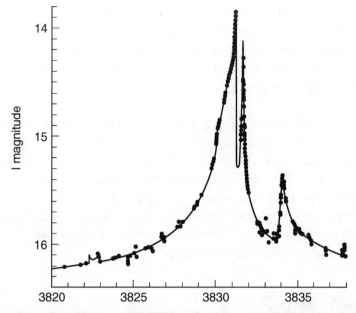

33. **The micro-lensing event OGLE-2006-BLG-109. The brightness of the star measured at 12 observatories is plotted against time in units of a day. The gravitational fields of two planets generate extremely rapid brightness fluctuations. These data yield the mass of the star as 0.51 $M_\odot$ and the planets' masses as 231 $\mathcal{M}_\oplus$ and 86 $\mathcal{M}_\oplus$ similar to Jupiter and Saturn.**

refractive index that is very close to unity ($n - 1 \simeq 5 \times 10^{-9}$) so light rays are bent through only tiny angles on their way to us unless they pass quite close to the surface of the Sun, where $n - 1$ is $\sim$ 100 times larger. Deflection of the light from stars by the Sun's gravitational field shifts the position of each star on the sky, and these shifts are in principle detectable by comparing images of a given star field taken at different times of the year.

When Einstein's theory was still new and untested, stars could be observed close to the Sun only during a solar eclipse. Moreover, the required measurement was extremely challenging as even for a star that is seen at the edge of the Sun's disc, the predicted shift in position is only 1.75 arcsec and all the stars in a small field of view will be shifted to a similar extent—measuring the absolute positions of stars is very much harder than measuring the angles between neighbouring stars. Nonetheless, during the 1919 eclipse a team led by Arthur Eddington measured shifts that were consistent with Einstein's prediction.

From a spacecraft stars can be observed even close to the Sun, but space telescopes avoid doing this because their delicate detectors would be fried if by chance any part of the Sun's disc came into the field of view. The Gaia satellite, launched in December 2013, can measure the positions of stars to such exquisite precision ($\lesssim 0.00001$ arcsec) that allowance must be made for Einstein's shift over the whole sky. In fact, the analysis also has to take acount of the deflection of light by planets.

Shapiro delays

In the 1960s it became possible to bounce radar waves off planets. The idea was to measure the time it took a pulse to return to Earth and to compare it with the time predicted by a general-relativistic model of the solar system. Two difficulties with these early experiments are that (i) reflection off a planet does not occur instantaneously, but over a period that depends on the planet's shape, and (ii) radio waves do not travel through interplanetary space at precisely the speed of light because space contains a tenuous plasma, which shifts the refractive index from unity. Both of these problems could be eliminated by replacing the planets with spacecraft programmed to respond to an outgoing signal by returning a signal after a precisely known delay—the impact of intervening plasma can be determined by comparing results obtained using different transmission frequencies because the delay caused by the plasma is frequency-dependent.

These experiments directly probe the gravitational field within the solar system, which is expected to be a slightly perturbed form of Karl Schwarzschild's solution of Einstein's equations. The conclusion is that any difference between the true and predicted fields must be smaller than parts in a thousand.

Pulsars and gravitational waves

Many, perhaps all, neutron stars have magnetic fields sticking out of them, that sweep the surrounding space as the neutron star spins. The rotating magnetic field can cause a beam of radio waves to sweep through space very much as the rotating lantern of a lighthouse sweeps over the ocean. The periodic passages of a beam over the Earth generates the highly characteristic radio signal of a *pulsar*.

A neutron star spins in a very regular way because it's hard for anything in its environment to apply a significant torque to it. So precision measurements can be made by comparing the times at which pulses are received with the times at which we calculate they were emitted, given the steady rotation of the neutron star. The most interesting object from this point of view is the Hulse–Taylor pulsar (page 90). The distance between the stars varies between 0.75 and 3.15 million kilometres (for comparison the radius of the Sun is 0.70 million kilometres). One of the neutron stars is a pulsar and general relativity predicts quite a complex pattern of arrival times for its radio pulses because the distance each pulse has to travel to reach us is constantly changing, as is the effective refractive index (see equation (6.2)) of the space the pulse has to traverse as it moves through the intense gravitational field of the binary neutron star. The predicted pattern is in excellent agreement with the measurements.

Any theory of gravity that is consistent with the symmetry that Lorentz's transformation uncovered will predict that a binary star emits gravitational waves. Indeed, when the sources of a gravitational field move, the field will be updated to the new source positions much sooner near the sources than at distant locations, and the updating will be accomplished by waves in the field that spread out from the moving sources. The basic physics of radiation is the same for gravitational waves as it is for electromagnetic waves, so the key to effective radiation is for the source (antenna) to be not much smaller than the wavelength of the waves (cf. Chapter 2, 'Emission by gas'). In the case of a binary star this condition translates into the stars moving not much slower than the speed of light. So two neutron stars that are almost touching would radiate gravitational waves with high efficiency and lose energy on a timescale of a few orbital periods (a fraction of a second), while a binary star with a separation of $\sim 1\,\text{AU}$ and a period of a year is a dreadfully inefficient radiator of gravitational waves. One of the best radiators we know is the Hulse–Taylor pulsar. It isn't a terribly good radiator: its timescale for energy loss is $\sim 0.3\,\text{Myr}$, or ~ 340 million orbital periods. Nonetheless, the precision of the measurements of pulse arrivals is such that the change in the period caused by gravitational radiation has been measured and found to be in excellent agreement with the prediction of Einstein's theory.

At the time of writing gravitational waves have yet to be detected because it's hard to build an efficient antenna if the fundamental requirement is massive bodies moving near the speed of light. Detectors are being perfected in which light is passed to and fro down two evacuated tunnels 5 km in length that are at right angles to one another. Interference fringes are observed between light that has been up and down one tunnel and light that has travelled the other tunnel. When a gravitational wave passes over the system it changes the effective refractive index (see equation (6.2)) within the tunnels and thus shifts the interference fringes. The expected effect from any astronomical source is absolutely tiny, but within a few years it should be observed. This feat will arguably be the all-time toughest of experimental physics.

Chapter 7
Galaxies

When you look up at the night sky on a dark night, the points of light above you are overwhelmingly stars within our Galaxy—two or three of the brighter points will be planets, and in a very dark site you may be able to make out the faint smudges of the Andromeda nebula or, if you can see far enough south, the Magellanic Clouds. By contrast, most of the sources detected by state-of-the-art telescopes are galaxies. The Universe seems populated by galaxies in the same way that our Galaxy is populated by stars.

Galaxy morphology

To an excellent approximation a galaxy consists of a huge number of point masses that move freely in the gravitational field that they jointly generate. Some of these masses are stars, but most are thought to be elementary particles of a still unknown type, which together comprise *dark matter*: material we cannot see but detect through its gravity. Astrophysics is made much simpler by the fact that, despite their hugely discrepant masses, stars and dark-matter particles have the same equations of motion—they move non-relativistically in a common gravitational field. Their typical orbits are nonetheless different: dark-matter particles tend to be on more energetic orbits that take them further from the centre of the galaxy, and we think their orbits are less concentrated around the galaxy's equatorial plane than are those of the stars, many of which are confined to a thin disc containing that plane.

The fraction of a galaxy's mass that is contained in stars rather than dark matter varies considerably. In the least massive galaxies, *dwarf spheroidal galaxies*, less than 1 per cent of mass is contained in stars. Our own Galaxy belongs to the class of galaxies that have the largest mass fractions in stars, and this fraction is $\sim$ 5 per cent. So whatever type of galaxy you choose to consider, dark matter dominates the overall mass budget. However, in a galaxy such as ours stars dominate the mass budget within a few kiloparsecs of the centre, while dark matter dominates further out. Dwarf spheroidal galaxies, by contrast, are dominated by dark matter at all radii.

A fundamental property of the population of galaxies is the *galaxy luminosity function* plotted in Figure 34. This shows the number of galaxies per unit interval of the logarithm of luminosity $\log L$. We see that there are myriads of faint galaxies, and rather few luminous galaxies: at the left-hand, low-luminosity, side of Figure 34, the luminosity function falls as a straight line, and then near a characteristic luminosity L^* it turns strongly downward. The *Schechter luminosity* L^* coincides almost exactly with the luminosity of our Galaxy.

From the galaxy luminosity function you can ask the question, if I repeatedly pick a star at random from all the stars in the Universe, what will be the average of the luminosities of the galaxy of the chosen stars? The answer turns out to lie close to L^*, so it is probably no accident that this is the luminosity of our Galaxy: we expect there to be innumerable civilizations in the Universe that have addressed this question, and most of them are warmed by a star that lies in a galaxy like ours.

Decomposition into components

It's helpful to imagine that a galaxy comprises a few *components*. A conspicuous component of our Galaxy is the disc, within which

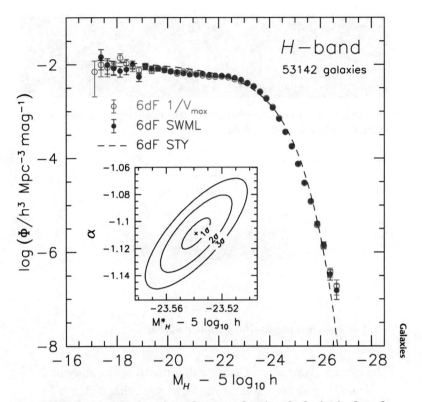

34. Galaxy luminosity function. The space density of galaxies is plotted against luminosity. Both scales are logarithmic.

the Sun lies (Figure 35). Stellar discs are found to have surface densities that fall off with radius roughly exponentially.

The stellar discs of galaxies frequently have embedded within them a gas disc like the Galaxy's gas disc, which we described in Chapter 2, 'The gas disc'. A gas disc is generally more radially extended than the stellar disc and distinctly thinner. Galaxies that possess a significant gas disc generally have spiral arms within both their gas and stellar discs. Spiral arms are generally absent when the gas disc is insignificant. Galaxies that have a prominent stellar disc but only an insignificant gas disc are called *lenticular* or *S0* galaxies. Our Galaxy is a *spiral* galaxy.

35. An image of our Galaxy constructed by counting half a billion stars. Clouds of obscuring dust are evident.

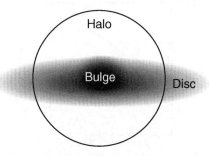

36. Cartoon showing disc, bulge/spheroid, halo.

The inner ~ 3 kpc of our Galaxy is dominated by the *bulge* or *spheroid* (Figure 36). As these names suggest, this stellar component is much less flattened towards the equatorial plane than the disc. The bulge of our Galaxy is not axisymmetric (which the disc nearly is), but forms a bar. The latter is about three times long as it is wide and its long axis lies in the Galactic plane. The bar rotates at the centre of our Galaxy like the beater of a food mixer. Meanwhile its stars circulate rather faster within the bar, moving on quite eccentric orbits. The bar excites spiral waves in the disc around it, but these waves move around the disc more slowly than the bar rotates.

In galaxies like ours the bulge is often barred. But not all bulges are barred, and not all spiral galaxies have bulges. For example, the third most luminous member of the *Local Group* of galaxies (of which our Galaxy is the second mast luminous member) is the *Triangulum Nebula* or *M33*, and it does not have a bulge. The existence of bulgeless galaxies such as M33 is currently rather puzzling for cosmologists.

Disc galaxies are ones in which the bulge is subordinate to the stellar disc (Figure 35). In an *elliptical galaxy*, by contrast, the bulge dominates the disc to the extent that the disc can be detected, if at all, only by the most minute quantitative analysis. Elliptical galaxies are usually, but not always, axisymmetric. In an elliptical galaxy the motions of stars are much less ordered than in a disc—where rotation around the approximate symmetry axis is very dominant—and rather less ordered than in the bulge of a galaxy such as ours.

Stellar dynamics

Since most of the mass of a galaxy is contained in particles, stars, and dark matter, which rarely collide, we need to understand how a large number of point masses move under their mutual gravitational attraction—a branch of astrophysics known as *stellar dynamics*. The particles in question can be stars or dark-matter particles, it makes little difference.

Is it safe to treat stars as point masses? The answer to this question is normally a resounding 'yes': about 3 Gyr from now our Galaxy will collide and merge with our nearest massive neighbour, the Andromeda Nebula. Then for hundreds of millions of years streams of 10^{11} stars from each galaxy will rush through each other as if in a stupendous military tattoo, and the number of physical impacts expected is less than one! In fact, the only environment in which treating stars as point particles may be problematic is in the immediate vicinity of a galaxy's central black hole. But even here physical contact can be significant only for giant stars, and such contact is likely merely to strip away the bloated atmosphere from such a star, without depriving it of most of its mass.

The Sun is on a near-circular orbit around the centre of our Galaxy some 8.3 kpc away. The gravitational pull towards the Galactic centre that holds it on this orbit is the sum of the pulls of the 10^{11} stars and zillions of dark-matter particles that make up the Galaxy. The fraction of this force contributed by the near neighbours of the Sun is negligible. This is a dramatically different situation from that in a solid or liquid: the force on an atom is completely dominated by the handful of nearest neighbours because the inter-atomic force declines with distance much more rapidly than the gravitational attraction between stars. Since the force on the Sun is dominated by large numbers of distant objects, the force would change very little if the Sun were moved a parsec or so in any direction, and it won't change very much in the next million years or so: the force is a very smooth function of position and time. Consequently, we can compute the orbit of a star such as the Sun to high precision by spreading the mass of each particle smoothly over a few inter-particle distances, and computing the gravitational field of the resulting continuous mass distribution. Our first step when considering any stellar system is to do just this and then to investigate the nature of orbits in the smooth gravitational field. In so far as this approach is adequate, we say the stellar system is *collisionless*.

You can specify an orbit by giving a position **x** and a velocity **v** that a star on this orbit has at some particular time. If every such pair (**x**, **v**) specified a different orbit, the space of orbits would be six dimensional, because a position and a velocity both live in three-dimensional spaces. But it is clear that different pairs (**x**, **v**) do not necessarily generate different orbits, because any pairs that occur at different times along an orbit obviously generate the same orbit.

Integrating orbits in typical galactic gravitational field reveals that the space of orbits is three-dimensional. That is, an orbit can be uniquely specified by three numbers. These numbers are called *constants of motion* because their values do not change as one moves along an orbit. The fundamental task of stellar dynamics is to learn how to compute suitable constants of motion J_i from a pair $(\mathbf{x}, \mathbf{v})$. That is, one needs an algorithm to compute three functions $J_i(\mathbf{x}, \mathbf{v})$.

The dynamical state of a galaxy can then be reduced to a density of stars and dark-matter particles in the imaginary space in which the Cartesian coordinates of points are the three numbers J_i. This is known as *action space*. Our knowledge of the action-space density of stars and dark-matter particles is still incomplete even for our own Galaxy, and is very incomplete for all external galaxies.

Galaxies, gases, and crystals

A galaxy, like a litre of gas or a diamond, consists of a large number of particles interacting with one another. Statistical physics provides a rather complete understanding of bottles of gas and crystals by starting from the concept of thermal equilibrium: the state into which the system relaxes if you leave it alone for long enough. The *principle of maximum entropy* tells us how to compute the arrangement of the system's particles when it's in thermal equilibrium. For example, in the case of a gas the principle of maximum entropy enables us to work out how many molecules are moving at any given velocity (the *Maxwellian distribution*), and what pressure the gas exerts given its energy and volume. Then by perturbing the state of thermal equilibrium, we determine the system's *transport coefficients*, such as its sound speed, thermal conductivity, viscosity, etc.

Unfortunately, the very first step in this chain of analysis is inapplicable to a galaxy, because a galaxy has no state of maximum entropy, and is therefore incapable of thermal equilibrium.

Entropy is disorder: the principle of maximum entropy simply says that in thermal equilibrium the system is as disordered as it can be given its energy, volume, and any other restrictions on rearrangements of its particles. A self-gravitating system such as a galaxy or a star can always increase its entropy by moving mass inwards to increase the intensity of the gravitational field near the centre, and then transporting outwards the energy that is released by this local contraction, and giving it to peripheral particles: this energy increases the distance from the centre to which these particles can cruise so their disorder increases. In Chapter 3, 'Life after the main sequence' we saw that late in the life of a star, its core contracts and its envelope swells up. This is because the star is increasing its entropy by the process we have just described.

Equilibrium dynamical models

The major problem we face, because galaxies can't attain thermal equilibrium, is how to deduce the basic distribution of stars and dark matter particles. Once we know what that is, we can compute the transport coefficients. But we we need to know what configuration to perturb, and we have no principle from which to deduce it. One workaround is to rely on a cosmological simulation of galaxy formation, and another is to fit a dynamical model to observational data.

Cosmological simulations do not provide useful predictions on their own because we lack the resources required to simulate the extremely complex physics of star and galaxy formation. Consequently, all simulations rely on mathematical formulae that it is hoped approximate the outcome of physics that has been left out, and the parameters in these formulae have to be calibrated against observations. So if you want to make galaxy models, you should go straight to the observations rather that bothering with simulations.

Elliptical galaxies   Elliptical galaxies are the easiest to model and significant numbers of such galaxies have been modelled dynamically. One important conclusion from these models is that most of these objects are nearly axisymmetric and flattened by their spins. However, the most massive elliptical galaxies spin very slowly and have triaxial shapes, rather like the kernel of plum. The low spins and triaxial shapes of these objects probably arise because they are the products of mergers of two gas-poor galaxies of comparable mass.

Another important conclusion from models of elliptical and lenticular galaxies is that the more luminous a galaxy is, the richer it is in heavy elements and the bigger is the proportion of its mass that is contributed by dark matter. The heavy-element richness of massive galaxies probably arises because supernovae have greater difficulty driving the products of nucleosynthesis out of the deeper gravitational potential wells of higher mass galaxies. The increasing contribution of dark matter may arise because more massive galaxies typically have lower star densities than less massive galaxies.

Models of the very centres of elliptical and lenticular galaxies are especially interesting because they may enable us to detect a central *super-massive black hole*. The key idea is that interior to the radius r_{infl} at which the black hole contributes as much to the gravitational field as do the stars, the random velocities of stars must rise roughly like $1/\sqrt{r}$. A secure detection of a black hole requires measurements of both the star density and the random velocities of stars very close to the centre. The Hubble Space Telescopy has been invaluable in gathering these data.

The inferred mass of the black hole proves to be tightly correlated with the magnitude of the random velocities of stars at radii that are significantly bigger than r_{infl}. This finding suggests a causal connection between the growth of the black hole and the growth of the galaxy's stellar population, and it has been argued that this is surprising because the stellar population is enormously more massive and physically extended than the black hole. However, the space density of quasars—black holes that are rapidly accreting cold gas—peaks at precisely the redshift $z \sim 2$ at which the cosmic star-formation rate was highest. Since stars form from cold gas, it seems likely that the growth rate of a black hole and its host stellar population both track the availability of cold gas, so it is natural that their current masses are tightly correlated.

Spiral galaxies The spiral galaxy that has been most extensively modelled is our own. Idealised equilibrium dynamical models of the thin and thick discs have been constructed, and from their vertical structure you can infer that 56 per cent of the gravitational force that holds the Sun in its orbit is generated by dark matter so only 44 per cent is generated by stars. The mass of the disc is consistent with the disc consisting exclusively of stars and gas rather than dark matter.

Slow drift

No star stays on the same orbit for the entire life of our Galaxy because the smooth, time-independent gravitational field that we assume when computing the constants of motion J_i is an idealization. The real gravitational field fluctuates around the idealized field for several reasons. First the disc supports spiral structure that is not reflected in our idealization. Second, the disc contains massive clouds of molecular gas that form, move through the disc and disperse in a random way. The gravitational fields of these objects we likewise ignored in our idealization. Third, no galaxy is isolated; other galaxies, many of them small, are constantly falling into a massive galaxy such as ours, and these objects can orbit through the galaxy for a long time before they disperse. In our idealization, we ignored these massive moving lumps. Finally, there are 'Poisson fluctuations' in the number of point masses in any volume: if the density of particles is such that on average we expect N objects in a volume V, then the number actually in that volume will fluctuate over time by $\sim \sqrt{N}$. Since the force arising from volume V is proportional to the number of masses it contains, the gravitational force will fluctuate by a fraction $\sqrt{N}/N = 1/\sqrt{N}$ of itself.

If we imagine a star or dark-matter particle to be on an orbit through the idealized smoothed gravitational field, we must consider it to be jiggled by a small random field. On account of this random field a star or dark-matter particle that's initially on the orbit **J** has a probability in a given time t of transferring to a different orbit **J**′. This is closely analogous to the Brownian motion of pollen grains on the surface of water: these are seen to jiggle around at random, so in some short time t they have a probability to move from position **x** to some nearby position **x**′. The net result of these random moves of individual pollen grains is to cause the density of pollen grains to *diffuse* through space: if initially the pollen grains are concentrated at **x**, over time they will spread out from **x** as they diffuse through space. In just the same way stars and dark-matter particles diffuse through action space.

Diffusion is particularly important for the stars of a stellar disc like the one we inhabit, because stars form in a very localized region of action space—the line associated with circular orbits in the disc's symmetry plane. As stars diffuse away from this line, their orbits become more eccentric and more highly inclined to the symmetry plane. Consequently, the random velocities of stars increase. Since the random velocities of gas molecules are associated with heat, we say the galactic disc 'heats'. The analogy is imperfect, however, since nothing heats the disc in the sense of supplying energy to it; it heats spontaneously and of its own accord, drawing the energy required for increased random velocities from its gravitational potential energy. The fluctuations in the gravitational field that cause disc heating are mostly generated by spiral structure (discussed below), but giant molecular clouds also contribute significantly. It is not clear whether infalling dwarf galaxies are significant contributors.

Globular clusters (Chapter 3, 'Globular star clusters') are very much like tiny galaxies. Their cores shrink and envelopes swell on account of the Poisson, $\sqrt{N}$, fluctuations in star density discussed above. In galaxies the timescale for Poisson fluctuations to cause measurable contraction in the core is much longer than the age of the Universe.

Destroying clusters   Most stars are born in small clusters—containing less than 1,000 $M_\odot$. Poisson fluctuations are large in such clusters and redistribute energy between the cluster's stars on a relatively short timescale. In any cluster a star requires only a finite amount of kinetic energy to escape completely from the cluster, and in any exchange of energy between stars there is always a chance that the gainer will acquire enough energy to escape. A star that does escape, never returns to risk losing energy. Since the energy of the cluster plus its escapees is conserved, the removal of positive energy by escapees must be mirrored by the energy of the surviving cluster becoming more negative. This is essentially the physics of evaporative cooling, which is what makes one shiver when wet in a draught.

As a cluster shrinks through evaporation of stars, Poisson fluctuations grow even more important and the rate of evaporative loss does not decrease even though the cluster's stars are on average becoming more tightly bound. Ultimately the cluster shrinks to a binary star; the energy released in the formation of this binary star has enabled every other star to escape to infinity.

We have just described what would happen to a small cluster if it were left alone for a long time. Actually, clusters are not isolated, but orbit through a galaxy and we will see on page 131 that they are gradually pulled apart by the galaxy's gravitational field. In fact, the Sun and every other star that is not now in a cluster has probably escaped from a cluster.

Spiral structure

Galaxies like ours contain spiral arms. In Chapter 4 we saw that the physics of accretion discs around stars and black holes is all about the outward transport of angular momentum, and that moving angular momentum outwards heats a disc. Outward transport of angular momentum is similarly important for galactic discs. In Chapter 4 we saw that in a gaseous accretion disc angular momentum is primarily transported by the magnetic field. In a stellar disc, this job has to be done by the gravitational field because stars only interact gravitationally. Spiral structure provides the gravitational field needed to transport angular momentum outwards.

In addition to carrying angular momentum out through the stellar disc, spiral arms regularly shock interstellar gas, causing it to become denser, and a fraction of it to collapse into new stars. For this reason, spiral structure is most easily traced in the distribution of young stars, especially massive, luminous stars, because all massive stars are young. In their short lives these stars don't stray far from their places of birth in an interstellar shock, so they trace the thin lines of the shocks. The gravitational field that caused the shock is mostly generated by numerous older, less massive stars. Their distribution forms a smoother spiral with broad arms, which is most prominent when a spiral galaxy is imaged in infrared light.

Spiral arms are waves of enhanced star density that propagate through a stellar disc rather as sound waves propagate through air. Like sound waves they carry energy, and this energy is eventually converted from the ordered form it takes in the wave to the kinetic energy of randomly moving stars. That is, spiral arms heat the stellar disc. Whereas sound waves heat air wherever they travel, spiral arms heat the disc at specific radii, where stars resonate with the wave. Exchange of energy between waves and particles at very specific locations is characteristic of collisionless systems, and is also crucial for the dynamics of electric plasmas. Our understanding of wave-particle interactions is still incomplete and the exact role they play in spiral structure and the evolution of galaxy morphology is controversial.

Origin of the bulge

Observations of the line-of-sight velocities of several thousand stars in the bulge/bar are well reproduced by bars that form in N-body simulations of self-gravitating discs. Bar formation proves to be a two-step process: quite a flat bar forms first, and a little later the bar experiences an instability which makes it vertically thicker. Thus the data are consistent with the proposition that all the stars of the bulge/bar and the disc were formed in the thin gas layer that occupies the equatorial plane. This accounts for the overwhelming majority of our Galaxy's stars: stars of the stellar halo were probably not formed in the Galactic plane, but less than 1 per cent of the Galaxy's stars belong to the stellar halo.

In N-body simulations of bars embedded in dark halos, the bar transfers angular momentum to the dark halo, with the consequence that the rate at which the figure of the bar rotates slows, and the bar grows stronger. Several pieces of evidence point to our bar having quite a fast rate of rotation, which is consistent with the N-body models if there has been a significant flux of gas through the disc and into the bar. In fact gas moves inwards because it loses angular momentum as it streams through spiral arms. As it approaches the bar's corotation radius, it starts to acquire angular momentum from the bar's rotating gravitational field, so its inward drift slows and its density increases—this is the origin of the *giant molecular ring*, a gas-rich region about 5 kpc in radius that surrounds the bar and dominates the Galaxy's star formation. Gas that escapes from the giant molecular ring and enters the bar rapidly loses angular momentum in shocks that we see in both numerical simulations and as dust lanes in external galaxies. After plunging rapidly through the bar, the gas builds up in a nuclear disc of radius ~ 0.2 kpc called the *central molecular zone*. There it feeds vigorous star formation as is evidenced by numerous supernova remnants in and near this disc.

Cannibalism

The Universe consists of dark-matter halos within which gravity has reversed the cosmic expansion, and our Galaxy is one of three substantial galaxies in one such halo, the Local Group. Because dwarf galaxies form in much greater abundance than giants, most of any halo's galaxies are dwarfs.

At the edge of any halo there are dwarf galaxies that are poised between their historic expansion from the halo's centre, and infall into the halo. These objects are on highly eccentric orbits in the halo's gravitational field and are just now at their most distant from the halo's centre. Several gigayears from now, they will come quite close to the centre of the halo. How will this experience affect them?

As a galaxy falls into a halo from r_{apo}, it attracts dark-matter particles that are near its path. To grasp the consequences of this attraction, it's easiest to imagine that the dark-matter particles constitute a fluid. As the dwarf passes, the fluid receives an impulse towards its line of flight and converges on that line (Figure 37). However it takes time for the fluid to move and its density to be enhanced along the line of flight. So the region of enhanced dark-matter density lies behind the dwarf. Hence the gravitational attraction from the region of enhanced density pulls the dwarf backwards; gravity acts to retard the dwarf's motion as if it were experiencing friction. In fact this phenomenon is called *dynamical friction*.

Since the impulse from the dwarf that creates the region of dark-matter overdensity is proportional to the mass of the dwarf, the mass of the overdensity is proportional to the mass of the dwarf. The drag on the dwarf, being proportional to the product of the masses of the dwarf and the overdensity is thus proportional to the *square* of the dwarf's mass, and the dwarf's deceleration is

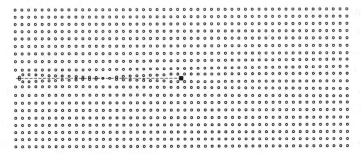

37. Formation of a wake behind a massive body. The massive body marked by the central black square is moving from left to right. Particles initially stationary at the nodes of a regular grid move towards it. A region of enhanced density of these particles is evident along the massive body's past track.

proportional to the dwarf's mass. Hence the more massive an infalling galaxy is, the more rapidly dynamical friction modifies its orbit.

If the dwarf did not experience dynamical friction, it would pass near the centre of the halo and move back out to nearly the radius from which it started; it wouldn't actually get to r_{apo} because other infalling material would have increased the mass interior to r_{apo} since it was last there and it wouldn't have enough energy to reach r_{apo}. If the dwarf has experienced much dynamical friction, it will stop moving out long before it reaches r_{apo}. Dynamical friction will cause each radius of turnaround to be smaller than the preceding one, and ultimately by this reckoning the dwarf will end up at rest in the centre of the halo. The halo, or the galaxy at its centre, will have cannibalized the dwarf.

If events unfolded precisely as just described, a dwarf of mass m that fell into a halo of mass $M(r)$ would survive for a time of order

$$t_{\text{eat}} = \frac{M(r_{\text{apo}})}{m} T,$$

where T is the period of a circular orbit at radius r_{apo}. If we extrapolate this formula to the case of a merger of equals, $m \simeq M(r)$, such as the future encounter of our Galaxy with the Andromeda Nebula, M31, we find $t_{eat} = T$, the initial orbital period. N-body simulations show that this extrapolated result is actually correct.

The dwarf that is currently closest to our Galaxy is the Sagittarius Dwarf, which lies $\sim$ 13 kpc from the Galactic Centre on the side opposite the Sun. The period T of a circular orbit at this radius is $T \simeq 0.5$ Gyr or $\sim 1/25$ of the current age of the Universe. Hence for t_{eat} to be smaller than the current age of the Universe, the Dwarf's mass must exceed 10^{10} M$_\odot$ since $M(20 \text{ kpc}) \simeq 2.5 \times 10^{11}$. The mass of the Dwarf is poorly determined, but it is much smaller than 10^{10} M$_\odot$, so we might conclude that it isn't going to be eaten any time soon. Not so! For we have underestimated our Galaxy's digestive powers.

Washed away on the tide

When an extended body orbits in a gravitational field, it is stretched along the line that joins its centre to the centre of attraction. The reason is that material on its surface that is closest to the centre of attraction feels a stronger gravitational force than material that is on the far side from the centre of attraction (Figure 38). Hence these two bodies of material want to accelerate at different rates. But they are parts of a single body, so they are obliged to accelerate at a single, compromise rate. This rate is too small for the material on the side of the centre of attraction, and too fast for the opposite material. The body responds to this discord by stretching along the line of centres so its gravitational field pulls back material nearest the centre of attraction, and urges on material on the opposite side. Since the physics we have described is how the Moon raises tides on the surface of the oceans, we say bodies are stretched by the *tidal field* of the body they are orbiting.

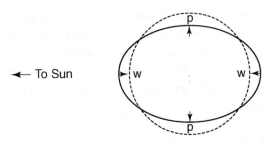

38. Tides without a moon: at the points of the Earth nearest to and furthest from the Sun the surface of the ocean (full curve) is higher than in the mean (shown dotted), so pressure is too small to balance the gravitational attraction of the Earth and net force w on a small volume of water is downwards. On the left this net force cancels a part of the gravitational attraction of the Sun, while on the right it adds to the Sun's attraction. These additions to the attraction of the Sun ensure that elements of the ocean at both locations orbit the Sun at the same angular velocity.

As dwarfs move inwards, their dark-matter halos are stretched to the point that particles are pulled clean out of the dwarf's clutches and start orbiting independently through the Galaxy's gravitational field. Particles that leave the dwarf on the same side as the centre of attraction transfer to orbits that have less angular momentum than the dwarf, and pull ahead of it, whereas particles that leave on the far side transfer to orbits of greater angular momentum and fall behind. In this way two *tidal tails* form and the dwarf grows less massive (Figure 39).

Now that the dwarf has become less massive, its gravitational field is weaker, and the points at which particles break free of the dwarf and start orbiting independently, edge towards the dwarf's centre. A vicious circle of mass loss proceeds, with the dwarf becoming less and less massive and the tidal tails growing longer and longer. At some stage the points at which particles break free come close enough to the centre that large numbers of stars as well as dark-matter particles stream out into the tidal tails. The Sagittarius dwarf reached this stage some time ago, and tidal tails

39. A simulation of tidal tails forming. The curve shows the orbit of the centre of the cluster that is being ripped apart.

containing its stars now wrap around the Galaxy at least once and perhaps twice.

By measuring the colours and brightnesses of millions of stars, it has been possible to examine the density of stars in spherical shells centred on the Sun. The density proves to be full of ridges and overdense patches. In fact, the ridges and overdense patches contain at least half the stars of the stellar halo. So it is probable that the stellar halo is entirely built up by tidal streams stripped from dwarf galaxies and globular clusters. When some of these objects first fell into our Galaxy, they may have been sufficiently massive to experience significant dynamical friction. But as they orbited, tides caused them to lose more and more of their dark-matter halos, and at some point their masses fell below the threshold for significant dynamical friction. However, tidal stripping continued relentlessly, and they were eventually completely digested without reaching the Galactic centre.

Chemical evolution

Nearly all the elements heavier than lithium have been produced since the Galaxy formed, and while they were formed, star formation has continued in the Galaxy. So in stars of various ages we have a fossil record of how the heavy-element content of the interstellar medium has evolved. Moreover, since old stars tend to have larger random velocities, the chemical content of a star is correlated with its kinematics.

It is hard to determine the age of an individual star because to do so one needs precise values for its mass, luminosity, and chemical content, and the mass is especially hard to determine. So astronomers like to use chemical content, which is easier to measure, as a surrogate for age.

Supernovae are the most important contributors to the enrichment of the interstellar medium with heavy elements. In Chapter 3, 'Exploding stars', we saw that there are two fundamentally different kinds of supernova: core collapse supernovae, which mark the deaths of massive stars ($M > 8\,\mathrm{M}_\odot$), and deflagration supernovae, which occur when a white dwarf accretes too much material from a companion. Since massive stars have short lives, a burst of core-collapse supernovae follows soon after (~ 10 Myr) an episode of star formation, while evolution to a white dwarf followed by significant accretion probably takes ~ 1 Gyr. So for the first gigayear of the Galaxy's life only core-collapse supernovae enriched the interstellar medium.

As we saw in Chapter 3, 'Exploding stars', deflagration supernovae produce mostly iron while core-collapse supernovae produce a wide spectrum of heavy elements. It follows that the abundance in the interstellar medium of iron relative to say magnesium or calcium would have been lower in the first gigayear of the Galaxy's life than it is now. Figure 40 shows that one can identify two populations of stars near the Sun: in the *alpha-enhanced* population the abundance of Mg and Ca is higher at a given value of the Fe abundance than it is in the *normal-abundance* population. We infer that stars of the alpha-enhanced population were formed in the first ~ 1 Gyr of the galaxy's life.

Within each population a wide range of abundances of Fe relative to hydrogen occurs. One possibility is that all stars with low values

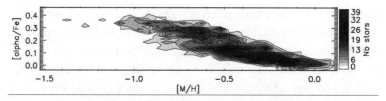

40. Stars formed in the first gigayear have low abundances of Fe relative to Mg and lie high up in this diagram. The horizontal coordinate is the abundance of iron. The contours show the density of stars in this plane.

of Fe/H formed before stars with higher values. But this conclusion is in general false because it is likely that early on interstellar gas was converted into stars more rapidly towards the centre of the Galaxy than further out, with the consequence that the heavy-element abundance rose near the centre much more rapidly than further out. Hence a given value of Fe/H was achieved earlier at small radii than at large radii, and a star with low Fe/H might have formed near the Galactic centre early in the life of the Galaxy, or more recently at large radii.

A star's ratio value of Mg/Fe provides a degree of discrimination between these possibilities: if Mg/Fe is high, the star must have formed in the first gigayear of the galaxy's life, and at that time only low values of Fe/H would have been attained at large radii, so a star with high Mg/Fe and high or moderate Fe/H must have formed quite far in.

Alpha-enhanced stars have large random velocities and extend further from the Galactic plane than normal-abundance stars. The data are consistent with the thick disc comprising alpha-enhanced stars and the thin disc normal-abundance stars.

Given the tight connections between time and place of birth, chemical composition, and present kinematics of disc stars, it is very useful to model chemical and dynamical evolution together. In such a model stars are born at some rate at each radius from the local interstellar gas, and enrich this gas with Mg and Ca shortly after they are formed, and with Fe a gigayear or so later. Stars are born on nearly circular orbits and gradually wander to more eccentric and inclined orbits. The interstellar gas at each radius is depleted by star formation and the ejection as a galactic wind of gas heated by supernovae. It is augmented by accretion of intergalactic gas. Spiral structure drives the star-forming gas slowly inwards, carrying heavy elements with it. The goal of such a model is to reproduce the observed kinematics of stars as a function of position, and the observed correlations between chemical composition and kinematics. This is currently an active area of research.

The great reservoir

In Chapter 2, 'The gas disc', we described our Galaxy's gas disc, which is typical of the gas discs of spiral galaxies and contains $\sim 6 \times 10^9$ $M_\odot$. Since the Galaxy turns gas into stars at a rate of about 2 $M_\odot$ yr^{-1}, this stock of material for star formation will be exhausted within ~ 3 Gyr. Is the current gas disc the small remnant of a massive gas disc from which the Galaxy has formed $\sim 5 \times 10^{10}$ $M_\odot$ of stars, or is it just a buffer between star formation and gas accretion?

Observations of other galaxies show that the star-formation rate in a disc is proportional to the surface density of cold gas, so in the absence of accretion, the mass of the gas disc decreases exponentially with time. Given the rate at which our Galaxy is currently forming stars and it current gas mass, it is easy to show that in the absence of accretion the gas mass 10 Gyr in the past would have been $\sim 1.7 \times 10^{11}$ $M_\odot$, and nearly all this mass would now be in stars. This absurd conclusion shows that our premise, that the Galaxy does not accrete gas, is false.

From the cosmic microwave background (CMB) we can read off the current cosmic mean density of ordinary matter. By measuring the luminosities of galaxies we know the cosmic mean luminosity density, so we can determine what mass of ordinary matter is required, on the average, to generate a given amount of luminosity – ~ $40\,M_\odot/L_\odot$. Now if you take any reasonably representative group of galaxies, from the group's luminosity, you can deduce the quantity of ordinary matter it should contain. This quantity proves to be roughly ten times the amount of ordinary matter that's in the galaxies. So most ordinary matter must lie *between* the galaxies rather than within them.

Intergalactic atomic hydrogen can be detected by its absorption of *Lyman alpha* photons that come to us from distant quasars (Chapter 4, 'Quasars'). Since light from the most distant quasars has taken most of the life of the Universe to reach us, it has sampled intergalactic space at essentially every cosmic epoch. Hence, measurements of the Lyman alpha line enable us to track the cosmic density of hydrogen over time. In the first gigayear, the density of hydrogen was roughly the same as the density of ordinary matter inferred from the CMB, but as time went on the density of hydrogen fell and now it is less than 1 per cent of the expected mean density of ordinary matter. Searches for 21 cm emission from atomic hydrogen between nearby galaxies confirms that there is very little intergalactic hydrogen now.

The natural interpretation of this finding is that the cosmic stock of hydrogen has been used up to create stars and galaxies, but studies of nearby galaxies show that they don't contain sufficient ordinary matter to make this a viable explanation. The accepted explanation is that the missing matter *is* in intergalactic space, but it so hot that it is completely ionized and can therefore not be detected with any spectral line of hydrogen.

Actually, it is natural for intergalactic gas to be extremely hot because the pressure within the gas can resist the gravitational pull of galaxies only if the temperature of the gas exceeds the *virial temperature*, which is the temperature at which the thermal velocities of atoms are comparable to the orbital speeds of dark-matter particles – in our Galaxy this temperature is $\sim 2 \times 10^6$ K but it varies as you move around the Universe, approaching 10^8 K in the richest clusters of galaxies. At the virial temperature the pressure in the gas provides an effective counterbalance to gravity, so the density of the gas tends to track the density of dark matter. Gas in which atoms move much slower than the dark-matter particles has too little pressure to resist gravity, and in equilibrium must be confined to a thin rotating disc.

Gas at the virial temperature emits X-rays. In rich clusters of galaxies the emitted X-rays are strong enough to be detected and indicate that the predicted quantity of gas is present (Figure 41). Outside rich clusters the X-ray emission is expected to be too weak and too soft to be detected by current telescopes. However, hints of the presence of the expected gas are found in the ultraviolet spectra of some objects.

Looking for absorption lines in the spectrum of a background object is by far the most sensitive way to search for gas (page 137).

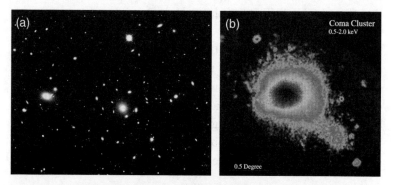

41. The Coma cluster, left (a) in optical light and right (b) in X-rays. The optical picture is only 0.32° wide while the X-ray image is 2.7° wide, so the optical picture shows only the centre of the cluster. The bright object at the top and to the right of centre in the optical picture is a star in our Galaxy; all other objects are galaxies.

The best waveband to examine for absorption lines depends on the temperature of the gas you are seeking, because the waveband should include photons that have the right energy to lift an ion out of its ground state into an excited state, and at high temperatures only tightly bound electrons remain bound to ions, and the ions can only be excited by energetic photons.

The ideal waveband to search for gas at more than 10^6 K is the X-ray band. Unfortunately X-ray telescopes are hard to build because X-rays tend to knock electrons out of mirrors rather than being reflected by them. Moreover for a given luminosity the rate of emission of X-ray photons is a thousand times less than the rate of emission of optical photons, so X-ray photons are scarce and statistical noise is a big problem. For these reasons sensitive searches for X-ray absorption lines are not possible. The Hubble Space Telescope has searched for absorption lines in ultraviolet spectra, however. Lines indicating the presence of ions such as five times ionized oxygen, O^{5+}, and three times ionized carbon, C^{3+} are detected in the spectra of quasars. The velocities of these lines indicate that the absorption occurs where the line of sight passes by a galaxy, but often at a great distance, typically 100 kpc.

The detected ions are common in gas that is cooler ($\sim 3 \times 10^5$ K) than we think the bulk of the gas is, and the usual interpretation of these data is that we are seeing absorption at the interface between the gas that fills most of intergalactic space and small clouds of much cooler ($\sim 10^4$ K) gas. This is a very active area of research and our interpretation of the data may be different in a few years.

Drivers of morphology

The nature of a galaxy is largely determined by three numbers: its luminosity, its bulge-to-disc ratio, and the ratio of its mass of cold gas to the mass in stars. Since stars form from cold gas, this last ratio determines how youthful the galaxy's stellar population is.

A youthful stellar population contains massive stars, which are short-lived, luminous, and blue (Figure 3). An old stellar population contains only low-mass, faint, and red stars. Moreover, the spatial distribution of young stars can be very lumpy because the stars have not had time to be spread around the system, just as cream just poured into coffee is distributed in blobs and streamers that quickly disappear when the cup is stirred. Old stars, by contrast, are smoothly distributed. Hence a galaxy with a young stellar population looks very different from one with an old population: it is more lumpy/streaky, bluer, and has a higher luminosity than a galaxy of similar stellar mass with an old stellar population.

The ratio of bulge-to-disc mass obviously affects the shape of the galaxy, especially when the disc is viewed edge-on. It also affects the compactness of the galaxy because a bulge of a given luminosity tends to be more compact than the corresponding disc.

Finally, the structure of a galaxy is profoundly affected by its luminosity, because the latter is related to its stellar mass, which is in turn connected to the speeds at which its stars and dark-matter particles move. Hence luminosity is connected to its virial temperature. A massive, luminous galaxy has a high virial temperature, which makes it hard for supernovae to drive gas out of the galaxy. Conversely, it is easy for supernovae to drive gas out of a low-mass, low-luminosity galaxy. Since today's cold gas is tomorrow's stars, driving gas out of a galaxy early on depresses the ratio of stars to dark matter, and this is thought to be the reason why low-luminosity galaxies have high ratios of dark-matter mass to stellar mass.

Galaxies are also shaped by their environments. Dense environments are rich in elliptical and lenticular galaxies, while abnormally under-dense environments are rich in dwarf irregular galaxies. Spiral galaxies like our own tend to inhabit regions of intermediate density. Such regions are made up of patches about a megaparsec in diameter within which the cosmic expansion has been reversed by gravity, so now each region consists of a number of galaxies that are falling towards one another. Our Galaxy lies in just such a region, that of the Local Group of galaxies. A few galaxies within such a group are stationary with respect to the local intergalactic gas, and act as sinks for this gas as it gradually cools. By processes that are still not fully understood but probably involve dynamical interactions with clouds of cold gas shot up out of the disc by supernovae (Chapter 2, 'The gas disc'), virial-temperature gas cools onto the disc, augmenting the supply there of cold, star-forming gas. Hence these galaxies, which include the Milky Way and our neighbours M31 and M33, remain youthful.

Smaller galaxies orbit around and through each of these major galaxies. Such galaxies cannot replenish their cold gas by accreting virial-temperature gas because they are moving through that gas too fast. Hence star formation in these galaxies dies out when whatever initial stock of cold gas they had is exhausted. This is why the nearer a satellite galaxy lies to the centre of its host galaxy, the less likely we are to see it forming stars: the density of virial-temperature gas increases inwards, so galaxies that are close in experience the strongest winds as they move at roughly the sound speed through the virial-temperature gas. These strong winds (think of travelling in an open-topped Boeing 747) sweep the satellite's own gas away (Figure 42).

Galaxy clusters The densest environments are rich clusters of galaxies. These are regions about a few megaparsecs in size in which gravity reversed the cosmic expansion quite long ago, so the density of galaxies and the virial-temperature are both high. Because the virial-temperature gas is dense, its X-ray emission is rather intense and can be detected by X-ray telescopes out to significant distances from cluster centres (Figure 41). The virial-temperature gas of a cluster is unusually hot because the characteristic velocities and temperatures of cosmic structures

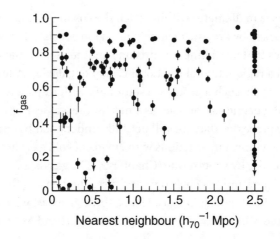

42. **The further a dwarf galaxy is from its neighbours, the more likely it is to contain little cold, star forming gas.**

increases with mass scale, and a rich cluster of galaxies is extremely massive ($\sim 10^{15}\,M_\odot$).

The virial temperatures of rich clusters exceed the temperature to which supernovae can heat interstellar gas. Hence no ordinary matter can have been blown out of these regions since the Big Bang, and their ratios of the mass of ordinary matter to dark matter should be the cosmic ratio. Within the experimental errors this is found to be the case.

Galaxies that orbit within the cluster cannot acquire intergalactic gas through cooling, so they are rarely forming significant numbers of stars. Galaxies that have cold, star-forming gas discs continually fall into rich clusters, and they go on forming stars at a declining rate until their stock of cold gas is exhausted. During this phase these are called *anaemic spiral galaxies*. Once they have ceased forming stars, anaemic spirals become lenticular galaxies—their stellar discs are fossil relics of their former star-forming gas discs.

Usually (but not always), the centre of a rich cluster has an exceptionally massive galaxy at its centre. Such a *cluster-dominant galaxy* is rather a special beast in that it is at rest with respect to the local intergalactic virial-temperature gas, so it can accrete gas and we might expect it to have a significant rate of star formation. Some of these galaxies, for instance NGC 1275 at the centre of the Perseus galaxy cluster, do have significant numbers of young stars, but most do not, and those that do lack dominant stellar discs.

Why do cluster-dominant galaxies not develop giant stellar discs that would make the parent galaxy a hugely scaled-up version of a spiral galaxy such as ours, in which the observed cluster-dominant galaxy would be the bulge and the cluster a halo of satellites? Astrophysicists do not have a complete answer to this question but the final answer will surely involve two key pieces of physics. First, the way a plasma cools changes fundamentally at around 10^6 K because the common elements carbon and oxygen are stripped of their last electrons around this temperature. Even a very low density of ions that have bound electrons greatly increases the cooling capacity of a plasma because a bound electron radiates photons *very* much more efficiently than a free electron. This phenomenon causes the time required for plasma of a given density to cool to rise steeply as the temperature increases from 10^6 K to 10^7 K, and this will hinder the formation of a disc of cool, star-forming gas around a cluster-dominant galaxy.

The second key fact is the presence of super-massive black holes (masses $\sim 10^9$ $M_\odot$) at the centres of most luminous galaxies. Any virial-temperature gas that does cool will do so near the black hole at the centre of the cluster-dominant galaxy, where the gas is under the most pressure and therefore densest. When the black hole accretes some of this gas, jets form (Chapter 4, 'Jets'), which blast through the surrounding virial-temperature gas, reheating it. Radio-frequency and X-ray observations of cluster-dominant galaxies provide clear evidence of this process. In particular, the X-ray spectra of several clusters show that the gas there is up to three times cooler than gas in the main body of the cluster, but the gas is not in the process of cooling right down. This finding is a clear sign that re-heating prevents the centre from settling into a steady state in which gas flows steadily onto the central body.

Our own Galaxy has a much smaller central black hole than a typical cluster-dominant galaxy: its mass is $4 \times 10^6 \, M_\odot$. Nonetheless, it has the capacity to re-heat the plasma that envelops it, and it probably regularly does so. At the moment it appears to be resting, like a dormant volcano. Its activity explains why the centre of our Galaxy is not the region of most intense star formation—that honour belongs to the central molecular zone. As we saw in our discussion of the Galactic bulge, cold gas is fed into this ~ 0.2 kpc radius zone from the ~ 5 kpc radius giant molecular ring. The key to understanding why every cluster of galaxies is not dominated by a giant spiral galaxy lies in understanding why these clusters have no analogues of the giant molecular ring, and this will likely be traced to the change around temperature $10^6 \, K$ in the way a plasma cools.

Chapter 8
The big picture

In Chapter 6 we discussed a number of phenomena that can be explained with the general theory of relativity. However, by far the biggest contribution of general relativity to astrophysics was to make it possible to discus the geometry and dynamics of the entire universe—it made cosmology a branch of physics rather than of philosophy or theology. We do not have space here for a systematic account of cosmology—for that the reader can turn to the *Cosmology: A Very short Introduction*. Instead we outline our current understanding of how stars and galaxies emerged from the big bang, in this way providing some context for the physical processes introduced in preceding chapters.

At heart cosmology is about three fluids: *dark energy*, which nobody understands; dark matter, which nobody can see; and the cosmic microwave radiation, which dominated the universe prior to a redshift $z \sim 3,000$ and can be studied in great detail as it constitutes the CMB. Dark energy became dominant rather recently ($z \sim 0.5$) and the intervening era was dominated by dark matter. From the cosmological standpoint what distinguishes these three fluids is the pressure they exert. Radiation exerts a positive pressure, dark matter exerts negligible pressure, and dark energy exerts negative pressure: that is it exerts tension.

According to general relativity, pressure is a source of gravitational attraction just as much as mass-energy. Hence the gravitational pull of the Sun on the Earth is larger than it otherwise would be because the pressure deep inside the Sun is high. Conversely, tension generates gravitational repulsion, and in the case of dark energy the repulsion generated by its tension overwhelms the attraction generated by its energy density. Since dark energy now dominates the universe, the latter is now being blasted apart by the gravitational repulsion that it generates. That this is happening was discovered by measuring the redshifts and distances of deflagration supernovae (Chapter 3, 'Exploding stars'). From these measurements the rate at which the universe was expanding at past epochs has been inferred, and it seems that around $z \sim 0.5$ the expansion rate started to increase, whereas previously it had decreased with time.

For the first 200,000 years after the big bang the universe was very nearly homogeneous and dominated by radiation, so gravity was strongly attractive and continuously slowed the expansion of the cosmic fireball. Because it exerts pressure, the radiation fluid did work on the expansion, and on account of doing this work its energy density diminished faster than the energy density of dark matter, which did negligible work because it exerted negligible pressure. Consequently, at redshift $z \sim 3{,}000$ the energy density of radiation fell below that of dark matter. At this point the initially tiny fluctuations in energy density with position began to grow at a significant rate, because the gravitational field was now predominantly generated by dark matter rather than radiation, and pressure did not oppose the tendency for some regions to become more dense than others. At this stage ordinary matter was tightly locked to the nearly homogeneous radiation fluid, so it did not participate in the clustering of dark matter. Then at a redshift $z \sim 1{,}000$ the temperature of the radiation dropped to the value at which electrons became bound to protons and alpha particles, to form hydrogen and helium atoms. The formation of these atoms effectively decoupled ordinary matter from the radiation fluid because the atoms scarcely scattered photons. Now nothing resisted the gravitational pull of ordinary matter into regions of enhanced dark-matter density, and the formation of structure got underway in earnest. We can study the *epoch of decoupling* in great detail by measuring the properties of the CMB because its constituent photons have travelled to us unmolested since the epoch of decoupling.

At the epoch of decoupling, the fluctuations in the density of dark matter were only parts in 100,000 so it took a long time for gravity to amplify them sufficiently for the regions of highest density to cease expanding and to collapse into stars and galaxies. This started to happen around redshift $z \sim 15$, and the first collapsed objects included massive stars. These stars radiated energetic photons that gradually re-ionized the hydrogen and helium atoms, a process that was largely complete by $z \sim 6$.

Currently we don't have much observational data relating to the redshift interval from $z \sim 1,000$ to 6, but from redshift 6 the observational record is significant. A few very massive galaxies must have already formed by then because luminous quasars are known at $z > 7$, and we know (Chapter 4, 'Quasars') that these are powered massive black holes that sit at the centres of massive galaxies.

Although at $z \sim 6$ there were a few massive galaxies, only a tiny fraction of the present-day stars had formed. At that time most galaxies were much smaller than present-day galaxies, and there was an abundance of cool, dense gas. As this gas flowed into nascent galaxies, it formed stars at a rate that continued to increase up to redshift $z \sim 2$. The flow of gas into galaxies was chaotic, so often the gas was not organized into a thin, flat disc like our Galaxy's present gas disc. Instead streams of gas raced hither and thither, crashing into each other and rapidly forming stars when they did so.

Massive black holes were in the thick of the melé, hoovering up gas as fast as they could. So galactic bulges and black holes grew fast at this time. Energy released by accretion onto the black holes was converted by the abundant ambient gas into optical and infrared photons, causing the region around each black hole to shine brightly as a quasar. Energy released at the deaths of massive stars heated the surrounding interstellar gas (Chapter 2, 'The gas disc'), with the consequence that an ever increasing fraction of the volume in and around galaxies became occupied by gas at the virial temperature or above. Gas hotter than the virial temperature flowed out into intergalactic space, carrying with it much of the heavy elements that had been synthesized by the recently deceased stars (Chapter 7, 'Drivers of morphology').

From redshift $z \sim 2$ the rate of star formation and black hole feeding gradually diminished as the flow of gas onto galaxies slackened, and more of the gas became too hot to form stars or to allow a black hole to gorge itself. In Chapter 7, 'Drivers of morphology', we described how these global trends influenced the morphologies of individual galaxies.

So there's a very brief history of the universe. Much of the physics involved is extremely complex and we are far from understanding how the various processes played out. Consequently, if we were to go into much more detail, we would soon reach the limits of our current understanding.

The universe is a huge canvas, and nature has wrought on it with very many techniques. Our knowledge of the canvas and of the artist's methods is growing rapidly, but we have much, much more to learn.

Further reading

Chapter 2: Gas between the stars

Bruce Draine, *Physics of the Interstellar and intergalactic Medium* (Princeton University Press, 2011). (A definitive, graduate-level text.)

T. Padmanabhan, *Theoretical Astrophysics*. Vol II: *Stars and Stellar Systems* (Cambridge University Press, 2001).

Chapter 3: Stars

Andrew King, *Stars: A Very Short Introduction* (Oxford University Press 2012).

T. Padmanabhan, *Theoretical Astrophysics*. Vol II: *Stars and Stellar Systems* (Cambridge University Press, 2001).

Dina Prialnik, *The Theory of Stellar Structure and Evolution* (Cambridge University Press, 2009). (A lucid, undergraduate-level text.)

A. Sarajedini, L.R. Bedin, B. Chaboyer, et al., The ACS Survey of Galactic Globular Clusters. I. Overview and Clusters without Previous Hubble Space Telescope Photometry, *The Astronomical Journal* 133, 4 (2007), 1658–72.

Chapter 4: Accretion

K.M. Blundell and M.G. Bowler, Letters, *Astrophysical Journal* 616 (2004), L159.

Juhan Frank and Andrew King, *Accretion Power in Astrophysics* (Cambridge University Press, 2002).

T. Padmanabhan, *Theoretical Astrophysics*. Vol II: *Stars and Stellar Systems* (Cambridge University Press, 2001).

Chapter 5: Planetary systems

Stephen Eales, *Planets and Planetary Systems* (Wiley-Blackwell, 2009).

W. Kley and R.P. Nelson, Annual Reviews, *Astronomy and Astrophysics*, 50 (2012), 211–49.

Chapter 6: Relativistic astrophysics

D.P. Bennett, S.H. Rhie, S. Nikolaev, et al., Masses and Orbital Constraints for the OGLE-2006-BLG-109Lb,c Jupiter/Saturn Analog Planetary System, *Astrophysics Journal* 713 (2010), 837–55.

M. Krause, Very Light Jets II: Bipolar Large Scale Simulations in King Atmospheres, *Astronomy and Astrophysics*, 431 (2005), 45–64.

Maurice van Putten and Amir Levison, *Relativistic Astrophysics of the Transient Universe* (Cambridge University Press, 2012).

Albert Einstein, *Relativity—The Special and General Theory* (Forgotten Books, 2015). (Get the story from the horse's mouth—this book is based on lectures Einstein delivered to engineers.)

P.A.M. Dirac, *General Theory of Relativity* (Princeton University Press, 2011). (A wonderfully short and elegant exposition by perhaps the second-greatest theoretical physicist of the 20th century.)

Chapter 7: Galaxies

James Binney and Michael Merrifield, *Galactic Astronomy* (Princeton University Press, 1998). (Graduate-level text.)

Bruce Draine, *Physics of the Interstellar and Intergalactic Medium* (Princeton University Press, 2011). (A definitive, graduate-level text.)

M. Geha, R. Blanton, and M. Masjedi, The Baryon Content of Extremely Low Mass Dwarf Galaxies, *The Astrophysical Journal*, 653 (2006), 240–54.

D.H. Jones, B.A. Peterson, M. Colless, and W. Saunders, Near-Infrared and Optical Luminosity Functions from the 6dF Galaxy Survey, *Monthly Notices of the Royal Astronomical Society* 369 (2006), 25–42.

A. Recio-Blanco, P. de Laverny, G. Kordopatis, et al., The Gaia-ESO Survey: The Glactic Thick to Thin Disc Transition, *Astronomy and Astrophysics* 567, id.A5 (2014), 21 pp.

J.L. Sanders and J. Binney, Stream-Orbit Misalignment I: The Dangers of Orbit-Fitting, *Monthly Notices of the Royal Astronomical Society* 433, 3 (2013), 1813–25.

L. Sparke and J.S. Gallagher, *Galaxies in the Universe* (Cambridge University Press, 2007). (Undergraduate-level text.)

"牛津通识读本"已出书目

古典哲学的趣味	福柯	地球
人生的意义	缤纷的语言学	记忆
文学理论入门	达达和超现实主义	法律
大众经济学	佛学概论	中国文学
历史之源	维特根斯坦与哲学	托克维尔
设计,无处不在	科学哲学	休谟
生活中的心理学	印度哲学祛魅	分子
政治的历史与边界	克尔凯郭尔	法国大革命
哲学的思与惑	科学革命	民族主义
资本主义	广告	科幻作品
美国总统制	数学	罗素
海德格尔	叔本华	美国政党与选举
我们时代的伦理学	笛卡尔	美国最高法院
卡夫卡是谁	基督教神学	纪录片
考古学的过去与未来	犹太人与犹太教	大萧条与罗斯福新政
天文学简史	现代日本	领导力
社会学的意识	罗兰·巴特	无神论
康德	马基雅维里	罗马共和国
尼采	全球经济史	美国国会
亚里士多德的世界	进化	民主
西方艺术新论	性存在	英格兰文学
全球化面面观	量子理论	现代主义
简明逻辑学	牛顿新传	网络
法哲学:价值与事实	国际移民	自闭症
政治哲学与幸福根基	哈贝马斯	德里达
选择理论	医学伦理	浪漫主义
后殖民主义与世界格局	黑格尔	批判理论

德国文学	儿童心理学	电影
戏剧	时装	俄罗斯文学
腐败	现代拉丁美洲文学	古典文学
医事法	卢梭	大数据
癌症	隐私	洛克
植物	电影音乐	幸福
法语文学	抑郁症	免疫系统
微观经济学	传染病	银行学
湖泊	希腊化时代	景观设计学
拜占庭	知识	神圣罗马帝国
司法心理学	环境伦理学	大流行病
发展	美国革命	亚历山大大帝
农业	元素周期表	气候
特洛伊战争	人口学	第二次世界大战
巴比伦尼亚	社会心理学	中世纪
河流	动物	工业革命
战争与技术	项目管理	传记
品牌学	美学	公共管理
数学简史	管理学	社会语言学
物理学	卫星	物质
行为经济学	国际法	学习
计算机科学	计算机	化学
儒家思想	亚当·斯密	天体物理学